236

Topics in Current Chemistry

Topics in Current Chemistry

Recently Published and Forthcoming Volumes

Natural Product Synthesis II
Volume Editor: Mulzer, J.H.
Vol. 244, 2004

Natural Product Synthesis I
Volume Editor: Mulzer, J.H.
Vol. 243, 2004

Immobilized Catalysts
Volume Editor: Kirschning, A.
Vol. 242, 2004

Transition Metal and Rare Earth Compounds III
Volume Editor: Yersin, H.
Vol. 241, 2004

The Chemistry of Pheromones and Other Semiochemicals II
Volume Editor:Schulz, S.
Vol. 240, 2004

The Chemistry of Pheromones and Other Semiochemicals I
Volume Editor:Schulz, S.
Vol. 239, 2004

Orotidine Monophosphate Decarboxylase
Volume Editors: Lee, J.K., Tantillo, D.J.
Vol. 238, 2004

Long-Range Charge Transfer in DNA II
Volume Editor: Schuster, G.B.
Vol. 237, 2004

Long-Range Charge Transfer in DNA I
Volume Editor: Schuster, G.B.
Vol. 236, 2004

Spin Crossover in Transition Metal Compounds III
Volume Editors: Gütlich, P., Goodwin, H.A.
Vol. 235, 2004

Spin Crossover in Transition Metal Compounds II
Volume Editors: Gütlich, P., Goodwin, H.A.
Vol. 234, 2004

Spin Crossover in Transition Metal Compounds I
Volume Editors: Gütlich, P., Goodwin, H.A.
Vol. 233, 2004

New Aspects in Phosphorus Chemistry IV
Volume Editor: Majoral, J.-P.
Vol. 232, 2004

Elemental Sulfur and Sulfur-Rich Compounds II
Volume Editor: Steudel, R.
Vol. 231, 2003

Elemental Sulfur and Sulfur-Rich Compounds I
Volume Editor: Steudel, R.
Vol. 230, 2003

New Aspects in Phosphorus Chemistry III
Volume Editor: Majoral, J.-P.
Vol. 229, 2003

Dendrimers V
Volume Editors: Schalley, C.A., Vögtle, F.
Vol. 228, 2003

Colloid Chemistry II
Volume Editor: Antonietti, M
Vol. 227, 2003

Colloid Chemistry I
Volume Editor: Antonietti, M
Vol. 226, 2003

Modern Mass Spectrometry
Volume Editor: Schalley, C.A.
Vol. 225, 2003

Hypervalent Iodine Chemistry
Volume Editor: Wirth, T
Vol. 224, 2003

Long-Range Charge Transfer in DNA I

Volume Editor: G. B. Schuster

With contributions by
D. Angelov · J.K. Barton · C. Behrens · J. Cadet
T. Carell · M.K. Cichon · T. Douki · B. Giese
F. Grolle · U. Hennecke · K. Kawai · T. Majima
U. Landman · F.D. Lewis · K. Nakatani
M.A. O'Neill · J.-L. Ravanat · I. Saito · G.B. Schuster
J.R. Wagner · M.R. Wasielewski

Springer

The series *Topics in Current Chemistry* presents critical reviews of the present and future trends in modern chemical research. The scope of coverage includes all areas of chemical science including the interfaces with related disciplines such as biology, medicine and materials science. The goal of eaxch thematic volume is to give the nonspecialist reader, whether at the university or in industry, a comprehensive overview of an area where new insights are emerging that are of interest to a larger scientific audience.

As a rule, contributions are specially commissioned. The editors and publishers will, however, always be pleased to receive suggestions and supplementary information. Papers are accepted for *Topics in Current Chemistry* in English.

In references *Topics in Current Chemistry* is abbreviated Top Curr Chem and is cited as a journal.

Visit the TCC home page at http://www.springerlink.com/

ISSN 0340-1022
ISBN 3-540-20127-0
DOI 10.1007/b84245
Springer-Verlag Berlin Heidelberg New York

Library of Congress Catalog Card Number 74-644622

Springer-Verlag is a part of Springer Science+Business Media

springeronline.com

Cover design: KünkelLopka, Heidelberg/design & production GmbH, Heidelberg
Typesetting: Stürtz AG, Würzburg

02/3020 ra – 5 4 3 2 1 0 – Printed on acid-free paper

Volume Editor

Dr. Gary B. Schuster

Georgia Institute of Technology
School of Chemistry and Biochemistry
Atlanta GA 30332-0400
USA
gary.schuster@cos.gatech.edu

Editorial Board

Topics in Current Chemistry also Available Electronically

For all customers who have a subscription to Topics in Current Chemistry, we offer the electronic version via SpringerLink free of charge. Please contact your librarian who can receive a password for free access to the full articles by registering at:

http://www.springerlink.com

If you do not have a subscription, you can still view the tables of contents of the volumes and the abstract of each article by going to the SpringerLink Homepage, clicking on "Browse by Online Libraries", then "Chemical Sciences", and finally choose Topics in Current Chemistry.

You will find information about the
- Editorial Board
- Aims and Scope
- Instructions for Authors
- Sample Contribution

at http://www.springeronline.com using the search function.

Preface

The central role played by DNA in cellular life guarantees a place of importance for the study of its chemical and physical properties. It did not take long after Watson and Crick described the now iconic double helix structure for a question to arise about the ability of DNA to transport electrical charge. It seemed apparent to the trained eye of the chemist or physicist that the array of neatly stacked aromatic bases might facilitate the movement of an electron (or hole) along the length of the polymer. It is now more than 40 years since the first experimental results were reported, and that question has been answered with certainty.

As you will earn by reading these volumes, Long-Range Charge Transfer in DNA I and II, today no one disputes the fact that charge introduced at one location in DNA can migrate and cause a reaction at a remote location. In the most thouroughly studied example, it is clear that a radical cation injected at a terminus of the DNA polymer can cause a reaction at a $(GG)_n$ sequence located hundreds of Ångstroms away.

In the last decade, an intense and successful investigation of this phenomenon has focused on its mechanism. The experimental facts discovered and the debate of their interpretation form large portions of these volumes. The views expressed come both from experimentalists, who have devised clever tests of each new hypothesis, and from theorists, who have applied these findings and refined the powerful theories of electron transfer reactions. Indeed, from a purely scientific view, the cooperative marriage of theory and experiment in this pursuit is a powerful outcome likely to outlast the recent intense interest in this field.

Is the quest over? No, not nearly so. The general agreement that charge can migrate in DNA is merely the conclusion of the first chapter. This hard-won understanding raises many important new questions. Some pertain to oxidative damage of DNA and mutations in the genome. Others are related to the possible use of the charge transfer ability of DNA in the emerging field of molecular-scale electronic devices. Still others are focused on the application of this phenomenon to the development of clinical assays.

It is my hope that these volumes will serve as a springboard for the next phase of this investigation. The foundation knowledge of this field contained within these pages should serve as a defining point of reference for all who explore its boundaries. For this, I must thank all of my coauthors for their effort, insight and cooperation.

Atlanta, January 2003 Gary B. Schuster

Contents

Contents of Volume 237
Long-Range Charge Transfer in DNA II

Volume Editor: Gary B. Schuster

ISBN 3-540-20131-9

Top Curr Chem (2004) 236:1–25
DOI 10.1007/b94409

Effects of Duplex Stability on Charge-Transfer Efficiency within DNA

Thierry Douki[1] · Jean-Luc Ravanat[1] · Dimitar Angelov[2] · J. Richard Wagner[3] · Jean Cadet[1]

[1] Laboratoire "Lésions des Acides Nucléiques", SCIB,
Département de Recherche Fondamentale sur la Matière Condensée, CEA/Grenoble, 38054 Grenoble Cedex 9, France
E-mail: jcadet@cea.fr

[2] Institute of Solid State Physics, Bulgarian Academy of Sciences, 1784 Sofia, Bulgaria

[3] Département de Médecine Nucléaire et Radiobiologie, Faculté de Médecine, Université de Sherbrooke, Sherbrooke, Québec, J1H 5N4, Canada

Abstract In the first part of the chapter, emphasis is placed on the description of the main reactions of radical cations of the four predominant DNA purine and pyrimidine bases and minor 5-methylcytosine in aerated aqueous solutions. Information is also provided on the final one-electron oxidation products of 8-oxo-7,8-dihydroguanine, an ubiquitous decomposition product of DNA with most chemical and physical oxidizing agents, that exhibits a low ionization potential, and is therefore an excellent sink for positive hole migration within double-stranded DNA. In the second part of the review, it is shown that duplex stability plays a major role in the redistribution of positive holes generated by high intensity UV laser pulses on purine and pyrimidine bases towards guanine residues. These results were obtained by measurement of several oxidized nucleosides within DNA using a sensitive and accurate high performance liquid chromatography-tandem mass spectrometry assay.

Keywords Radical cations · Oxidized DNA bases · Biphotonic laser photochemistry · Oxidative DNA damage distribution

Abbreviations

dAdo	2′-deoxyadenosine
dCyd	2′-deoxycytidine
dGuo	2′-deoxyguanosine
5-MedCyd	5-methyl-2′-deoxycytidine
Thd	thymidine
FapyAde	4,6-diamino-5-formamidopyrimidine
FapyGua	2,6-diamino-4-hydroxy-5-formamidopyrimidine
5-FordUrd	5-formyl-2′-deoxyuridine
5-HmdUrd	5-(hydroxymethyl)-2′-deoxyuridine
5-OHdCyd	5-hydroxy-2′-deoxycytidine
5-OHdUrd	5-hydroxy-2′-deoxyuridine
8-oxodAdo	8-oxo-7,8-dihydro-2′-deoxyadenosine
8-oxodGuo	8-oxo-7,8-dihydro-2′-deoxyguanosine
ThdGly	5,6-dihydroxy-5,6-dihydrothymidine
Sp	spiroiminodihydantoin
Gh	guanidinohydantoin
MQ	menadione

1 Introduction

Oxidation reactions, which are ubiquitous within cells, have been shown to be implicated in aging, several types of cancer, and various other diseases, such as arteriosclerosis, arthritis, cataract and diabetes [1–4]. Major sources of endogenous oxidation processes include aerobic metabolism leading to the leakage of superoxide radicals from mitochondria and endoplasmic reticulum, and inflammation involving the production of reactive species by specific enzymes such as NADPH oxidase and myeloperoxidase [5]. Exogenous chemical and physical agents, including environmental carcinogens, ionizing radiation and UV-A light may also contribute to the induction of oxidative reactions. In this respect, DNA, together with other key biomolecules that include membrane lipids and proteins, appear to be critical cellular targets.

Much attention has been paid during the two last decades to the elucidation of oxidation reactions mediated by oxygen ($^{\cdot}OH$ radical, H_2O_2, $^1O_2$, O_3,

HOCl) and nitrogen ($NO^{\cdot}$, $^{-}OONO$) reactive species that give rise to a wide array of DNA modifications [for early reviews see, 6–9]. In contrast, interest in the one-electron oxidation of DNA is more recent; however, a notable exception concerns detailed investigations of the early events associated with the direct effect of ionizing radiation [for a recent review, see 10]. Major attention devoted to charge transfer reactions within DNA during the last five years has provided a strong impetus to this field of research. As it will be discussed extensively in several chapters, most of the approaches have involved the specific generation of radical cations either at the C-4 of 2-deoxyribose or at proximal guanines within defined sequences of DNA fragments, and the investigation of hole transfer processes to distant guanine residues. Interestingly, this has led to major achievements with the formulation of several mechanisms of transfer, such as multistep hopping, phonon-assisted polaron-like hopping, and coherent super-exchange [11–15].

A different strategy has been applied in our work, that emphasizes the importance of DNA stability on hole transfer within double-stranded DNA. This work is based on determination of the overall yield of oxidized nucleosides that arise from the conversion of initially generated purine and pyrimidine radical cations within DNA exposed to two-photon UVC laser pulses. On the one hand, this work benefits from the excellent current knowledge of chemical reactions involving the radical cations of DNA bases, and on the other hand, from major analytical improvements that include recent availability of the powerful technique of high performance liquid chromatography-electrospray ionization-tandem mass spectrometry (CLHP-ESI-MS/MS) [16–18].

2
One-Electron Oxidation Reactions of the Pyrimidine and Purine DNA Bases

Relevant information on the main pyrimidine and purine radicals generated by interaction of ionizing radiation with DNA and model compounds was gained from extensive ESR studies in the solid state using dry samples and frozen aqueous solutions [19–22]. Evidence was provided that positive holes are located on guanine whereas electron attachment involves pyrimidine bases in γ- or X-irradiated DNA after warming. These results strongly imply the occurrence of significant electron transfer processes. This was confirmed through the identification and measurement of final degradation products resulting from the conversion of radiation-induced pyrimidine and purine radicals upon thawing the frozen solutions or dissolving the dry nucleoside samples into aqueous solutions [23–26].

However, the biochemical significance of the latter studies is challenged by the fact that the transformation of transient purine and pyrimidine radicals into diamagnetic decomposition products is oxygen-independent in the solid state. Therefore, it is necessary to study the chemistry of one-electron nucleobase intermediates in aerated aqueous solutions in order to investigate the role of oxygen in the course of reactions that give rise to oxidation products within DNA and model compounds. In this respect, type I photosensitization with 2-methyl-1,4-naphthoquinone (menadione) [27–29], and high intensity UV laser photolysis [30, 31] were found to efficiently ionize both purine and pyrimidine nucleobases despite the fact that guanine has the lowest oxidation potential among DNA components.

2.1 One-Electron Oxidation Reactions of Thymine DNA Base

An almost complete description of reactions resulting from the one-electron oxidation of the thymine moiety of DNA and related model compounds such as thymidine (**1** in the reaction scheme shown in Scheme 1) is now available. Photoexcited 2-methyl-1,4-naphthoquinone (MQ) has been used to efficiently generate pyrimidine radical cations **2** through one-electron oxidation of **1** [29, 32]. Two main degradation pathways that involve hydration and deprotonation of initially-generated thymidine radical cations (**2**) were proposed from isolation and characterization of the main oxidized nucleosides, including hydroperoxides and other stable decomposition products [33]. The exclusive formation of four *cis* and *trans* diastereomers of 5-hydroperoxy-6-hydroxy-5,6-dihydrothymidine (**5**) may be rationalized in terms of the specific generation of oxidizing 6-hydroxy-5,6-dihydrothymid-5-yl radicals (**3**) [34] through hydration at C-6 of thymidine radical cations (**2**). The formation of hydroperoxides (**5**,**11**) together with a number of well known stable oxidation products of thymidine (**6**–**8**) accounts for approximately 60% of the total product mixture arising from the hydration of thymidine radical cations [29, 32, 35, 36]. On the other hand, competitive deprotonation of thymidine radical cation (**2**) accounts for 40% and explains the formation of methyl oxidation products (**11**–**13**).

The formation of thymidine hydroperoxides probably involves fast addition of molecular oxygen to neutral 6-hydroxy-5,6-dihydrothymid-5-yl (**3**) and (2-deoxyuridyl) 5-methyl (**9**) radicals, with rates of reaction that are controlled by diffusion [32]. In subsequent steps, about half of the resulting peroxyl radicals **4**,**10** are reduced by superoxide radicals [28] leading to the formation of hydroperoxides **5**,**11**. These hydroperoxides undergo decomposition in aqueous solution with lifetimes that range (at 37 °C) from 8–10 h for the *trans* diastereomers **5**, 16–35 h for the *cis* diastereomers **5**, and up to about 8 weeks for the highly stable 5-(hydroperoxymethyl)-2′-deoxyuridine (**11**). The hydrolytic decomposition of thymidine 5-hydroxy-6-hydroperoxides predominantly gives rise to the (5R^*)- and (5S^*)-diastereomers of 1-(2-deoxy-*β*-D-*erythro*-pentofuranosyl)-5-hydroxy-5-methylhydantoin (**8**). Depending on the presence of metal ions, the latter hydroperoxide may also transform into the four *cis* and *trans* diastereomers of 5,6-dihydroxy-5,6-dihydrothymidine (ThdGly) (**7**) and *N*-(2-deoxy-*β*-D-*erythro*-pentofuranosyl) formamide (**6**). In comparison, transformation of 5-(hydroperoxymethyl)-2′-deoxyuridine (**11**) leads to the formation of 5-(hydroxymethyl)-2′-deoxyuridine (5-HMdUrd) (**12**) and to a much lesser extent, 5-formyl-2′-deoxyuridine (5-FordUrd) (**13**).

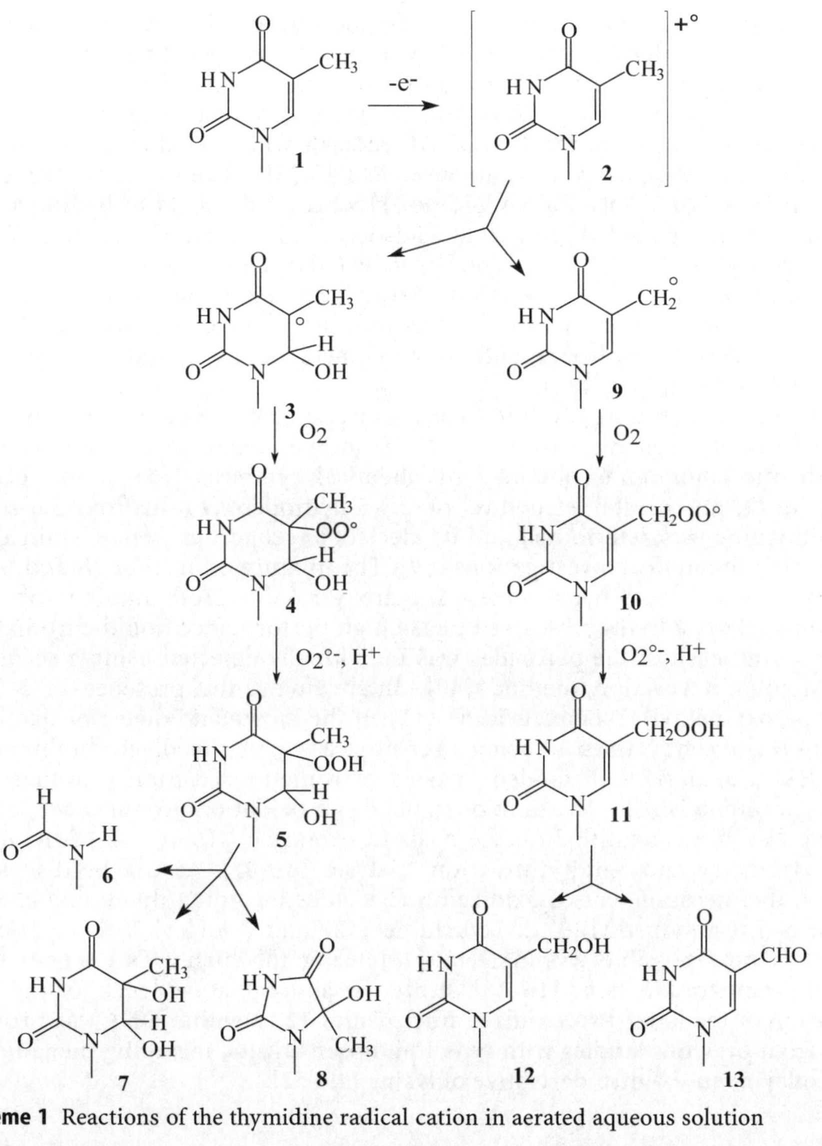

Scheme 1 Reactions of the thymidine radical cation in aerated aqueous solution

On the other hand, a number of thymidine oxidation products may be formed from dismutation reactions of initial peroxyl radicals (**4**,**10**) generated by the one-electron oxidation of dThd **1** by MQ-photosensitization in aerated aqueous solutions. These reactions give rise to highly reactive oxyl radicals [34] that can either undergo hydrogen atom abstraction to give thymidine glycols (**7**) or undergo β scission and 5,6-pyrimidine bond cleavage with subsequent loss of a pyruvyl group. Ring contraction of the resulting ureid may provide the (5R^*)- and (5S^*)-diastereomers of **8** whereas hydrolysis leads to the formation of formamide nucleoside **6**. Dismutation reactions of **10** would predominantly generate 5-formyl-2′-deoxyuridine (**13**).

Interestingly, one-electron oxidants partly mimic the effects of ˙OH radicals in their oxidizing reactions with the thymine moiety of nucleosides and DNA. In fact, the main reaction of ˙OH radicals with **1** is addition at C-5 that yields reducing radicals in about 60% yield [34, 38]. The yield of ˙OH radical addition at C-6 is 35% for thymidine (**1**) whereas the yield of hydrogen abstraction on the methyl group that leads to the formation of 5-methyl-(2′-deoxyuridylyl) radical (**9**) is a minor process (5%). Thus, the two major differences in terms of product analysis between the oxidation of dThd by one-electron oxidants and that by the ˙OH radical are the distribution of thymidine 5-hydroxy-6-hydroperoxide diastereomers and the overall percentage of methyl oxidation products.

The structure of thymidine (Thd) hydroperoxides **5,11** was assigned on the basis of extensive ^{1}H and ^{13}C-NMR measurements and comparison to authentic compounds obtained by chemical synthesis [33]. More recently, the X-ray crystal structure of *cis*-5-hydroperoxy-6-hydroxy-5,6-dihydrothymine was determined and its electronic properties were examined by *ab initio* theoretical investigations [39]. The mixture of the four *cis* and *trans* diastereomers of 5-hydroperoxy-6-hydroxy-5,6-dihydrothymidine (**5**) was completely resolved by reversed phase high performance liquid chromatography and each of the peroxides was individually detected using a sensitive post-column reaction method [40]. Interestingly, the presence of 5-(hydroperoxymethyl)-2′-deoxyuridine (**11**) in the enzymatic digest of oxidized DNA was recently assessed using a sensitive assay that involved combination of HPLC analysis with tandem mass spectrometry detection (Ravanat and Cadet, unpublished). The bulk of stable dThd oxidation products **6–8, 12,13** have also been identified on the basis of extensive ^{1}H and ^{13}C NMR, mass spectrometry and X-ray diffraction analyses [41–47]. As will be discussed later, the measurement of oxidized nucleosides including thymidine glycols (**7**), 5-(hydroxymethyl)-2′-deoxyuridine (**12**) and 5-formyl-2′-deoxyuridine (**13**) is now possible within DNA [48] using the highly and sensitive of HPLC-tandem MS assay [16–18]. It may be added that evidence for the formation of the latter two oxidized nucleosides **12,13** within DNA was provided from previous studies with type I photosensitizers, including menadione, riboflavin, and a nitro derivative of lysine [49–52].

2.2
One-Electron Oxidation Reactions of Cytosine and 5-methylcytosine DNA Base

2′-Deoxycytidine (dCyd) (**14** in Scheme 2) is also an excellent target for one-electron oxidation reactions mediated by triplet excited menadione. On the basis of extensive identification of dCyd photooxidation products, it was concluded that this nucleoside decomposes by competitive hydration and deprotonation reactions of cytosine radical cations with yields of 52% and 40%, respectively [53]. It was also found, on the basis of ^{18}O labeling experiments, that hydration of cytosine radical cations (**15**) predominantly occurs

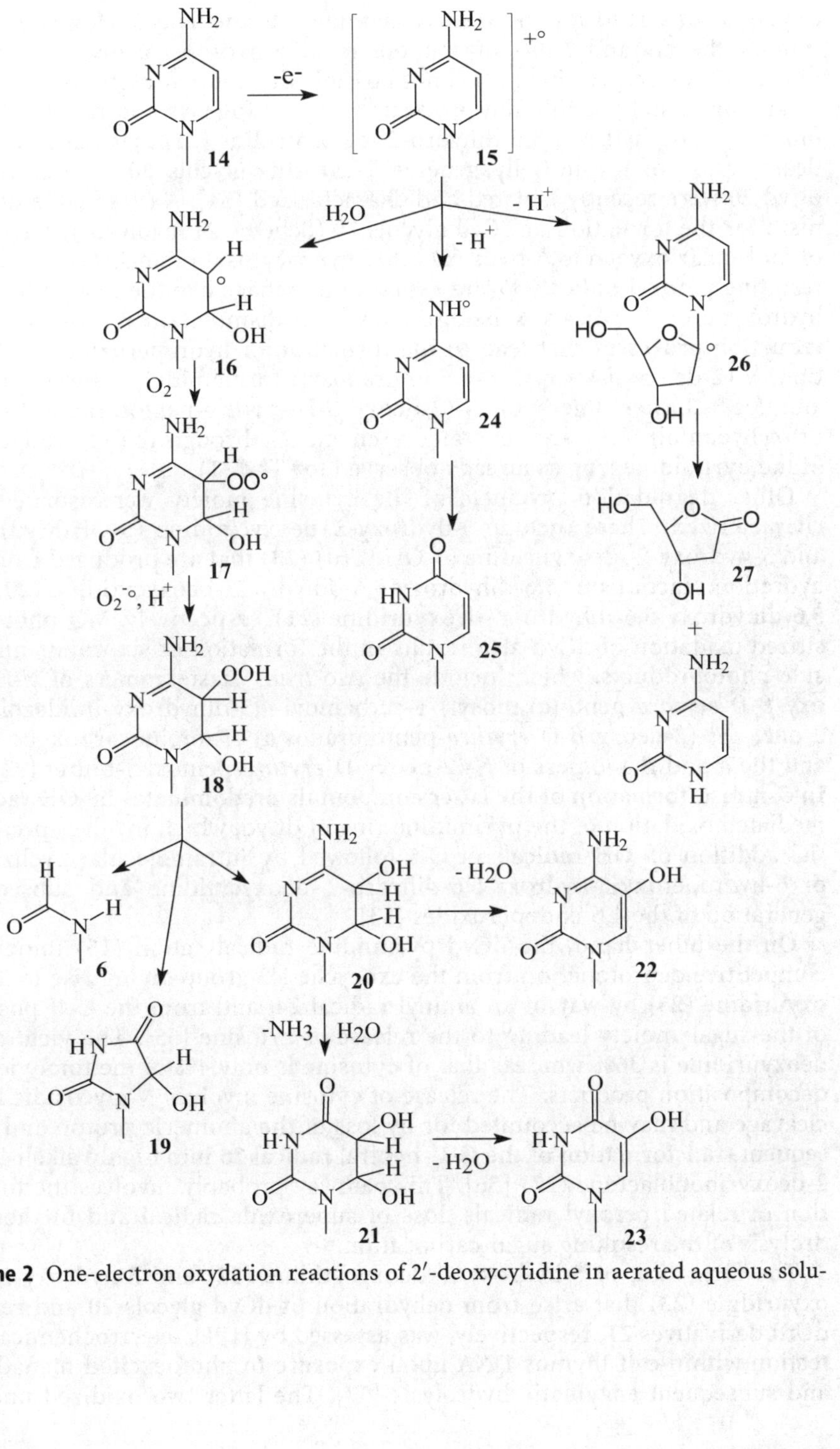

Scheme 2 One-electron oxydation reactions of 2′-deoxycytidine in aerated aqueous solutions

at C-6, as observed for the related thymine intermediate **2**. However, isolation of the *cis* and *trans* diastereomers of 5-hydroperoxy-6-hydroxy-5,6-dihydro-2′-deoxycytidine (**18**) is not possible due to their high instability.

The main stable oxidation products were the four *cis* and *trans* diastereomers of 5,6-dihydroxy-5,6-dihydro-2′-deoxyuridine (**21**) that arise from fast deamination of the initially generated cytosine glycols **20**. The glycols of dCyd **20** were recently isolated and characterized [54]. A reasonable mechanism for the formation of dCyd glycols **20** (Scheme 2) involves fast addition of molecular oxygen to 6-hydroxy-5,6-dihydrocytosyl radicals (**16**) [32]. The resulting peroxyl radicals **17** are expected to behave like the related thymine hydroxyperoxyl radicals **4**, being involved in dismutation reactions and/or reduction processes that lead to the formation of hydroperoxides. In addition, *N*-(2-deoxy-*β*-D-*erythro*-pentofuranosyl)-formamide (**6**) and the (5*R**)- and (5*S**)-diastereomers of *N*-(2-deoxy-*β*-D-*erythro*-pentofuranosyl)-5-hydroxyhydantoin (**19**) are generated (Scheme 2) through the rearrangement of the pyrimidine ring as already observed for Thd (**1**).

Other degradation products of the cytosine moiety were isolated and characterized. These include 5-hydroxy-2′-deoxycytidine (5-OHdCyd) (**22**) and 5-hydroxy-2′-deoxyuridine (5-OHdUrd) (**23**) that are produced from dehydration reactions of 5,6-dihydroxy-5,6-dihydro-2′-deoxycytidine (**20**) and 5,6-dihydroxy-5,6-dihydro-2′-deoxyuridine (**21**), respectively. MQ-photosensitized oxidation of dCyd also results in the formation of six minor nucleoside photoproducts, which include the two *trans* diastereomers of *N*-(2-deoxy-*β*-D-*erythro*-pentofuranosyl)-1-carbamoyl-4;5-dihydroxy-imidazolidin-2-one, N^1-(2-deoxy-*β*-D-*erythro*-pentofuranosyl)-N^4-ureidocarboxylic acid and the *α* and *β* anomers of *N*-(2-deoxy-D-*erythro*-pentosyl)-biuret [32, 53]. In contrast, formation of the latter compounds predominates in ˙OH radical-mediated oxidation of the pyrimidine ring of dCyd, which involves preferential addition of ˙OH radicals at C-5 followed by intramolecular cyclization of 6-hydroperoxy-5-hydroxy-5,6-dihydro-2′-deoxycytidine and subsequent generation of the 4,6-endoperoxides [53].

On the other hand, the dCyd pyrimidine radical cation (**15**) undergoes competitive deprotonation from the exocyclic N3 group giving rise to 2′-deoxyuridine (**25**) by way of an aminyl radical **24**, and from the C-1′ position of the sugar moiety leading to the release of cytosine [55]. The yield of 2′-deoxyuridine is 36% whereas that of cytosine is only 4% of the total yield of decomposition products. The release of cytosine involves *N*-glycosidic bond cleavage and may be accounted for by loss of the anomeric proton and subsequent transformation of the C-1′-neutral radical **26** into highly alkali-labile 2-deoxyribonolactone (**27**) [56]. This pathway probably involves the formation of related peroxyl radicals, loss of superoxide radical and further hydrolysis of the resulting sugar carbocation.

The formation of 5-hydroxy-2′-deoxycytidine (**22**) and 5-hydroxy-2′-deoxyuridine (**23**) that arise from dehydration of dCyd glycols **20** and related dUrd derivatives **21**, respectively, was assessed by HPLC-electrochemical detection within calf thymus DNA upon exposure to photoexcited menadione and subsequent enzymatic hydrolysis [57]. The latter two oxidized nucleo-

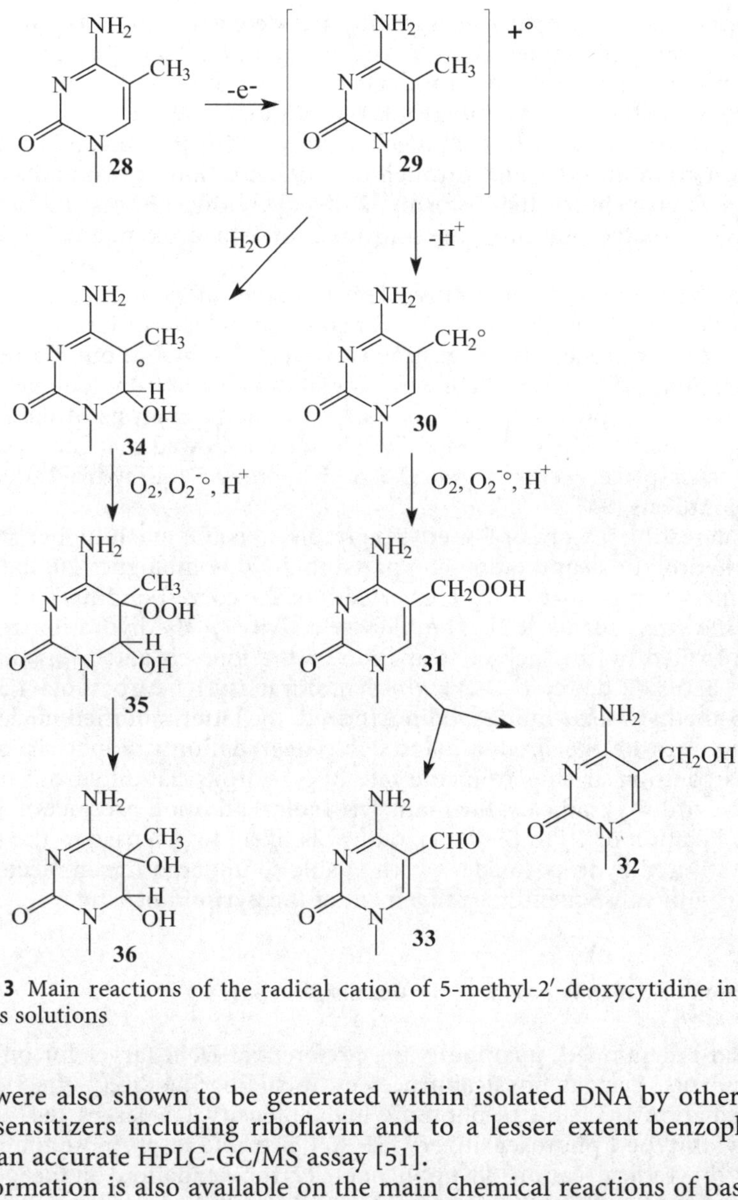

Scheme 3 Main reactions of the radical cation of 5-methyl-2′-deoxycytidine in aerated aqueous solutions

sides were also shown to be generated within isolated DNA by other type I photosensitizers including riboflavin and to a lesser extent benzophenone using an accurate HPLC-GC/MS assay [51].

Information is also available on the main chemical reactions of base radical cations (**29**) of 5-methyl-2′-deoxycytidine (5-MedCyd) (**28**), a minor DNA nucleoside that plays a major role in regulating gene expression at CpG sequences. Interestingly, deprotonation of 5-MedCyd radical cations readily occurs with 60% yield. The main reaction involves loss of a proton from the methyl group as previously observed for Thd. Subsequently, the (2-deoxycytidylyl)-5-methyl radical **30** incorporates oxygen to give rise to the corresponding peroxyl radical, which undergoes reduction by $O_2^{\cdot-}$ and then protonation to yield 5-(hydroperoxymethyl)-2′-deoxycytidine (**31**) [58]. The half-life of hydroperoxides at 24 °C was found to be 9.5±0.5 h. In addition, two stable methyl oxidation products were isolated and assigned as 5-(hydroxymethyl)-2′-deoxycytidine (**32**) and 5-formyl-2′-deoxycytidine (**33**). They likely arise from a Russell mechanism [59, 60] that involves the transient formation of tetroxides through the condensation of two peroxyl radicals. It was also shown that 5-formyl-2′-deoxycytidine (**33**) is the predominant MQ-mediated photooxidation product of 5-(hydroxymethyl)-2′-deoxycytidine (**32**).

It may be noted that competitive deprotonation of **29** at C-1′ gives rise to 2-deoxyribonolactone (**27**) with the concomitant release of free 5-methylcytosine as minor processes. Interestingly, competitive hydration of 5-MedCyd radical cations (**29**) occurs exclusively at C-6 as inferred from labeling experiments with $^{18}O_2$ (**36**) [61]. Thus, mass spectrometry analysis of the four *cis* and *trans* diastereomers of 5-MedCyd glycols **36** showed that incorporation of $^{18}O_2$ takes place exclusively at C-5 of 6-hydroxy-5,6-dihydro-2′-deoxycytyd-5-yl radicals (**34**).

An interesting feature of 5-MedCyd glycols **36** is the much higher stability toward hydrolytic deamination compared to dCyd homologues **20**, the *cis* diastereomers being more easily converted into the corresponding Thd glycols **7** than the *trans* forms [62]. The high selectivity of the hydration reaction was confirmed by the lack of formation of the four *cis* and *trans* diastereomers of 3-(2-deoxy-*β*-D-*erythro*-pentofuranosyl)-1-carbamoyl-4,5-dihydroxy-5-methyl-2-oxo-imidazolidine. Indeed, the latter modified nucleosides were found to be the predominant stable degradation products in aerated aqueous solution arising from the fate of 5-hydroxy-5,6-dihydro-5-methyl-2′-deoxycytid-6-yl radicals, the main $^{\cdot}OH$ radical addition product of 5-MedCyd **28**. Fixation of O_2 to the latter radical is likely to give rise to the related C-6 substituted hydroperoxide, which is able to undergo intramolecular cyclization with subsequent rearrangement of the pyrimidine ring.

2.3 One-Electron Oxidation Reactions of the Guanine DNA Base

As already mentioned, guanine is the preferential DNA target for one-electron oxidants. Various physical processes, including the direct effect of ionizing radiation [63] and bi-photonic high intensity UV lasers [64–66], together with type I photosensitizers [51, 67], are able to promote the formation of the radical cation **38** upon one-electron oxidation of the guanine moiety of DNA and related nucleosides such as 2′-deoxyguanosine (dGuo) (**37**). This is also the case for chemical agents that include Co (II) ion in the presence of benzoyl peroxide [68] peroxyl and oxyl radicals [69] and several radicals such as $CO_3^{\cdot-}$ [70] $Br_2^{\cdot-}$, $(SCN)_2^{\cdot-}$ [71] Tl^{2+} or $SO_4^{\cdot-}$ [72].

The two overwhelming oxidation products of the purine moiety of dGuo **37** arising from the transformation of guanine radical cations **38** were isolated and identified as 2,2-diamino-4-[(2-deoxy-*β*-D-*erythro*-pentofuranosyl)amino]-5(2*H*)-oxazolone (**41**) and its precursor 2-amino-5-[(2-deoxy-

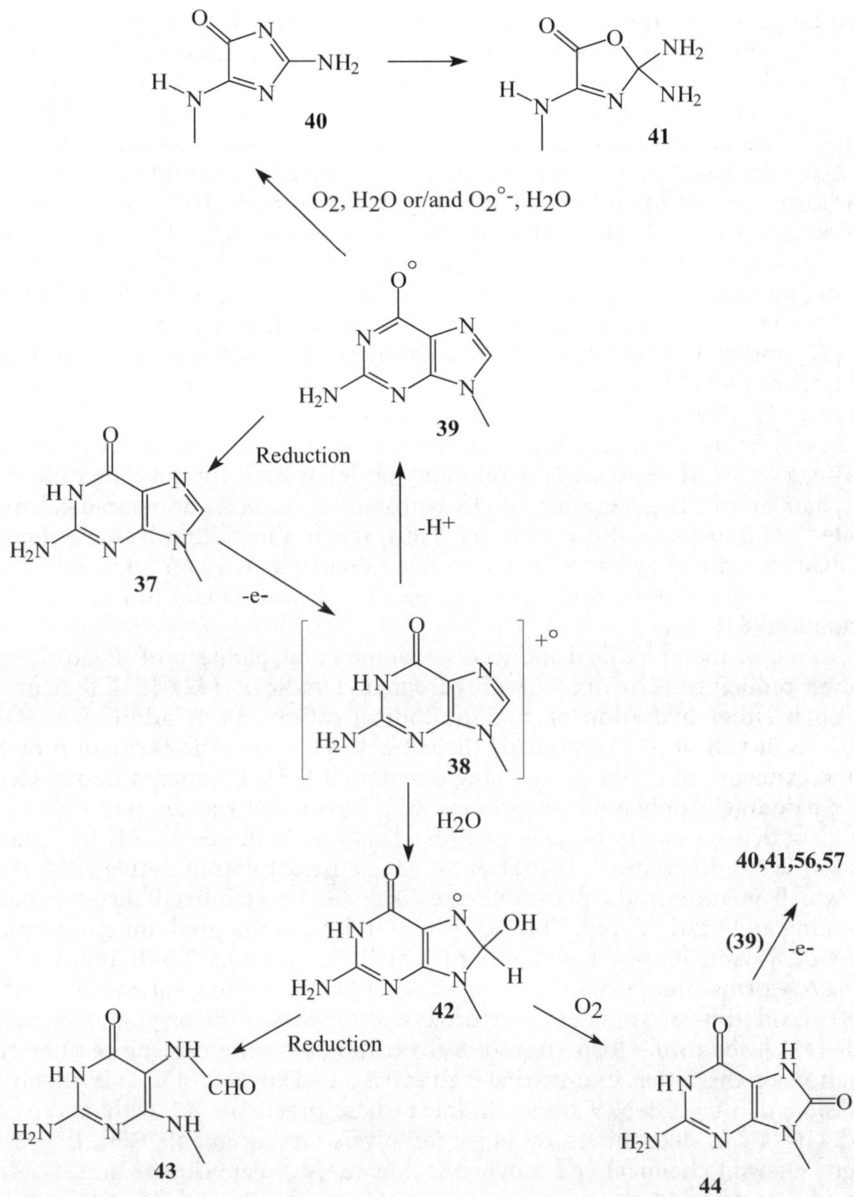

Scheme 4 One-electron oxidation reactions of 2′-deoxyguanosine in aerated aqueous solutions

β-D-*erythro*-pentofuranosyl)amino]-4H-imidazol-4-one (**40**) [73, 74]. The mechanism of their production (Scheme 4) may be rationalized in terms of the transient formation of oxidizing guanilyl radicals **39** that arise from deprotonation of **38** [75]. The resulting neutral radicals of dGuo(-H)· **39** may also be generated through dehydration from the overwhelming $^{\cdot}OH$ radical C-4 adduct of the guanine moiety (k=6×10^3 s^{-1}). In a subsequent step, **39** or tautomeric C-5 centered radicals are expected to react with O_2 or more likely $O_2^{\cdot-}$ with an estimated rate constant of 3×10^9 M^{-1} s^{-1} [75]. Then, addition of water is expected to take place at the 7,8-double bond of resulting peroxyl radicals, followed by rearrangement. The release of formamide [76] leads to the formation of oxazolone **41** through quantitative hydrolysis of unstable imidazolone **40** (half-life=10 h in aqueous solution at 20 °C) [73, 74]. However, side-oxidation of 8-oxo-7,8-dihydro-2′-deoxyguanosine (8-oxodGuo) **44** by highly oxidizing oxyl radicals **39** may also contribute to the formation of the oxazolone **41** and imidazolone **40** nucleosides. It may be noted that both **41** [77] and **40** [78] are strongly alkali-labile compounds whereas 8-oxodGuo **44** under similar hot piperidine treatment is stable in terms of DNA strand cleavage [77, 79].

Interestingly, the nucleophilic addition of water in the sequence of events giving rise to **41** represents a relevant model system for investigating the mechanism of the generation of DNA-protein cross-links under radical-mediated oxidative conditions [80, 81]. Thus, it was shown that lysine tethered to dGuo via the 5′-hydroxyl group is able to participate in an intramolecular cyclization reaction with the purine base at C-8, subsequent to one electron oxidation [81].

A second major radical-induced decomposition pathway of dGuo **37** involves reducing 8-hydroxy-7,8-dihydroguanyl radicals (**42**) [82] that arise through either hydration of guanine radical cations **38** or addition of $^{\cdot}OH$ radicals at C-8 of **37** (17% yield) (Scheme 4). However, the formation of **44** upon exposure of dGuo **37** and single stranded DNA fragments to one-electron oxidants is only a minor process [83]. This may be accounted for by an efficient deprotonation of guanine radical cations **38** at neutral pH [82] since its pK_a is 3.9. In contrast, hydration of **38** is efficient within native DNA [84] in which guanine radical cations are likely to be stabilized through base stacking and base pairing. This was found to lead to the predominant formation of 8-oxodGuo **44** together with small amounts of 2,6-diamino-4-hydroxy-5-formamidopyrimidine (FapyGua) (**43**) in aerated aqueous solution (98). Oxidation of transient 8-hydroxy-7,8-dihydro-2′-deoxyguanosyl radicals (**42**) leads to the formation of 8-oxodGuo (**44**), whereas competitive reduction of the latter intermediate gives rise to FapyGua **43**. This involves scission of the C-8–N-9 bond of the radical precursor **42** with a rate of k=2×10^5 s^{-1} as determined by pulse radiolysis measurements [75]. Interestingly, relevant chemical and biochemical features of FapyGua **43** and its adenine homologue **49** have recently become available [85–88]. By site-specifically inserting unstable FapyGua derivatives into defined sequence DNA fragments as dinucleoside monophosphates, it was found that this modification is a weakly piperidine-labile lesion [87].

Relevant kinetic information on two competitive reactions of guanine radical cations within double stranded DNA, namely hydration and hole transfer to another guanine residue, has been examined [13]. Thus, the pseudo-order rate for hydration of guanine radical cations **38** has been estimated to be 6×10^4 s^{-1} at neutral pH. This is lower by about two orders of magnitude than the rate of hole migration between single guanines separated by two AT base pairs [13]. The rate constant of guanine radical decay in DNA is estimated by direct absorption to be slower, in the order of 100 μs time-scale or $<1\times10^4$ s^{-1} [89].

2.4
One-Electron Oxidation Reactions of Adenine DNA Base

One-electron oxidation of the adenine moiety of DNA and 2′-deoxyadenosine (dAdo) (**45**) gives rise to related purine radical cations **46** that may undergo either hydration to generate 8-hydroxy-7,8-dihydroadenyl radicals (**47**) or deprotonation to give rise to the 6-aminyl radicals **50**. The formation of 8-oxo-7,8-dihydro-2′-deoxyadenosine (8-oxodAdo) (**48**) and 4,6-diamino-5-formamidopyrimidine (FapyAde) (**49**) is likely explained in terms of oxidation and reduction of 8-hydroxy-7,8-dihydroadenyl precursor radicals **47**, respectively [90]. Another modified nucleoside that was found to be generated upon type I mediated one-electron oxidation of **45** by photoexcited riboflavin and menadione is 2′-deoxyinosine (**51**) [29]. The latter nucleoside is likely to arise from deamination of 6-aminyl radicals (**50**). Overall, the yield of formation of 8-oxodAdo **48** and FapyAde **49** upon one-electron oxidation of DNA is about 10-fold-lower than that of 8-oxodGuo **44** and FapyGua **43**, similar to $^{\cdot}$OH radical mediated reactions [91].

Interestingly, it was recently shown that the one-electron oxidation reaction promoted by photoexcited menadione (MQ) gives rise to N^6-formyladenine (**52**) and N^6-acetyladenine (**53**) residues (Scheme 5) in several dinucleoside monophosphates including d(ApA), d(CpA) and d(ApC) [92].

2.5
One-Electron Oxidation of 8-oxodGuo

There is a growing body of evidence showing that 8-oxodGuo **44**, which has an oxidation potential about 0.5 eV lower than that of dGuo **37** [93], is a preferential target for numerous one-electron oxidizing agents. These include Na_2IrCl_6 [94, 95], $K_3Fe(CN)_6$, $CoCl_2/KHSO_5$ [96], high-valent chromium complex [97] peroxyl radicals [98], triplet ketones, oxyl radicals [99] ionizing radiation through the direct effect [100], and riboflavin as a type I photosensitizer [101].

Insights were recently gained into the kinetic features of the one-electron oxidation of 8-oxodGuo **44** by various inorganic and organic radicals including $Br_2^{\cdot-}$, $N_3^{\cdot}$, $SO_4^{\cdot-}$, $CH_3O_2^{\cdot}$, $CCl_3O_2^{\cdot}$, TyrO$^{\cdot}$, Trp$^{\cdot+}$ and dGuo(H)$^{\cdot}$ **39** [102]. The rate of reaction of the latter oxyl radicals **39** arising from deprotonation of the guanine radical cations **38** with 8-oxodGuo **44** was found to be 0.46×10^9 M^{-1} s^{-1} at pH 7.0. This fast reaction, together with the fact that the rate of O_2 addition to oxyl guanine radicals **39** is likely to be very low ($k<10^2$ M^{-1} s^{-1}), would explain why 8-oxodGuo **44** is barely detectable upon

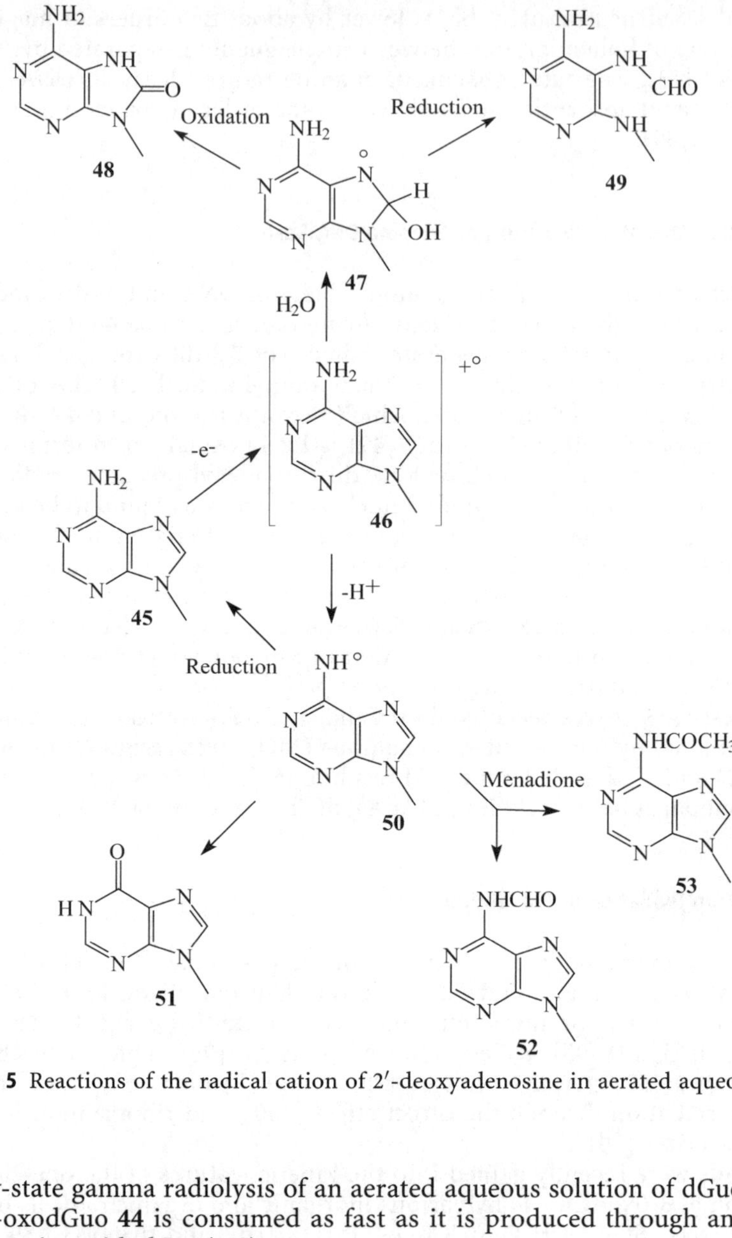

Scheme 5 Reactions of the radical cation of 2′-deoxyadenosine in aerated aqueous solutions

steady-state gamma radiolysis of an aerated aqueous solution of dGuo **37**. In fact 8-oxodGuo **44** is consumed as fast as it is produced through an oxidation reaction with dGuo(–H)˙ **39**.

Interestingly, the two (R^*)- and (S^*)-diastereomers of spiroiminodihydantoin (Sp) nucleosides **56** were found to be the predominant one-electron oxidation products of 8-oxodGuo **44** and 8-oxo-7,8-dihydroguanosine (8-oxoGuo) at neutral pH. The formation of (Sp) containing products **56** was rationalized by the transient generation of 5-hydroxy-8-oxo-7,8-dihydrogua-

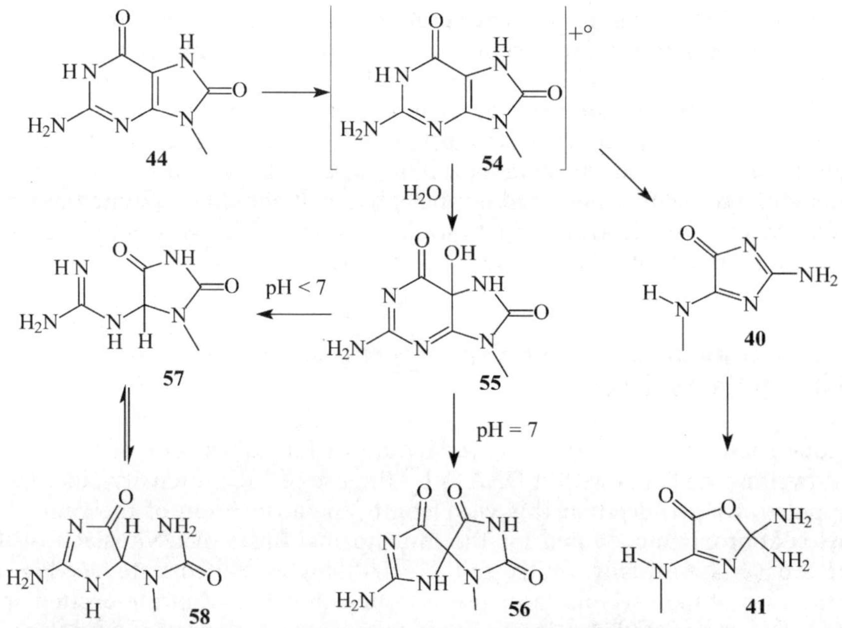

Scheme 6 One-electron oxidation of 8-oxo-7,8-dihydro-2′-deoxyguanosine in aerated aqueous solutions

nine derivatives (55) that rearrange via an acyl shift (Scheme 6). The latter precursors were found to undergo a different decomposition pathway under slightly acidic conditions; this involves opening of the 5,6-pyrimidine ring followed by a decarboxylation reaction and formation of the two diastereomers of guanidinohydantoin (Gh) derivatives **57**. It was suggested that the latter Gh nucleosides **57** exist in a dynamic equilibrium with a pair of diastereomeric iminoallantoin compounds **58** [95, 101]. The oxazolone nucleoside **41**, together with its imidazolone **40** precursor were also found to be one-electron oxidation products of 8-oxodGuo **44**, although they were generated in lower yields than (Sp) **56** and (Gh) **57** nucleosides [101].

Peroxynitrite is able to oxidize 8-oxoGuo through the intermediacy of $CO_3^{\cdot -}$ radicals generated in bicarbonate buffered aqueous solutions. This was found to lead to the predominant formation of (Sp) ribonucleosides in the presence of thiols [103]. Several secondary ONOO mediated oxidation products of 8-oxodGuo **44** were characterized within oligonucleotides including oxazolone **41**, oxaluric acid and cyanuric acid [104]. The two diastereomers of the (Sp) 2′-deoxyribonucleosides **56** were found to be the main reaction products of 8-oxodGuo **44** with either HOCl or a myeloperoxidase-H_2O_2-Cl^- system. The formation of Sp containing nucleosides **56** was accounted for by either initial Cl^+ addition or two electron oxidation of the base moiety of **44** [105]. The reactivity of 8-oxodGuo **44** residues toward various one-electron oxidation agents was investigated in defined DNA fragments that show different sequence context [106]. Interestingly, it was shown that 8-oxodGuo **44** residues, when stacked in a duplex with a 3′ neighboring guanine, were more susceptible to one-electron oxidation than any other possible doublet with the three other normal bases. This finding, which parallels the previously higher observed reactivity of the 5′ guanine in GG doublets toward one-electron oxidants, is in agreement with *ab initio* calculations of the ionization potential of nucleobases. It should be pointed out that in contrast to one-electron oxidation, the reaction of $^1O_2$ with 8-oxodGuo **44** does not show any marked sequence selectivity.

3 High Intensity UV Laser Photochemistry of Isolated DNA: Photophysical Features

A convenient way to induce the formation of radical cations of the purine and pyrimidine bases within DNA is by the use of high intensity 266 nm laser pulses [65]. Indeed, at this wavelength, the absorption of incident photons is approximately equal for the four normal bases of DNA. Each of the nucleobases are initially excited to their first singlet excited state (see Fig. 1). In the case of nanosecond laser pulses, one can expect that the excited species have time to undergo intersystem crossing to the corresponding triplet excited state. Due to the high intensity of UV laser pulses, the relatively long-lived triplet state likely absorbs a second photon. The energy absorbed by the molecule in this biphotonic process exceeds the ionization threshold of purine and pyrimidine bases by more than two eV, and as a result, it directly yields the radical cations of DNA bases [30]. This occurs roughly with the same efficiency for the four bases.

The formation of final oxidation products such as 8-oxodGuo **44**, 8-oxodAdo **48**, 5-HMdUrd **12** and 5-FordUrd **13** is monitored using the highly specific method of HPLC-MS/MS [17]. Application of this powerful technique as described in the next section has also been extended to the quantitative measurement of the main DNA photoproducts that arise from the reaction

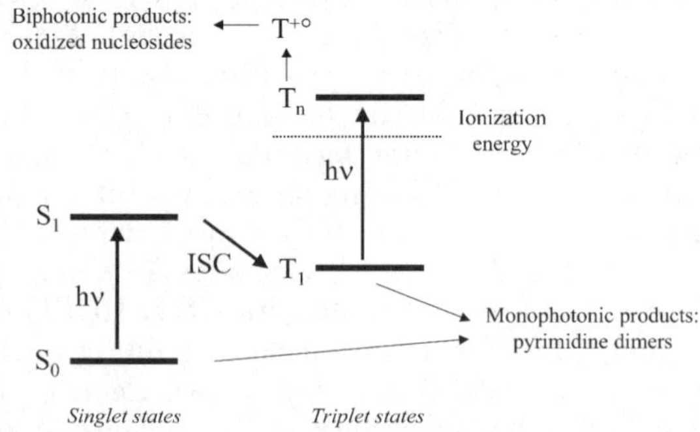

Fig. 1 Principles of UV laser-induced biphotonic ionization of DNA bases

of singlet and triplet excited pyrimidine bases with adjacent ground state thymine or cytosine bases. This leads to the formation of *cis-syn* cyclobutadipyrimidines together with pyrimidine (6–4) pyrimidone photoproducts and their Dewar valence isomers [29, 107, 108].

3.1 HPLC-MS/MS Measurement of One- and Two-Photon Induced DNA Base Modifications

A specific approach for the measurement of base damage to DNA involves the hydrolysis of DNA into monomeric units. Acidic hydrolysis leads to the release of bases while enzymatic treatment yields nucleosides. The resulting mixture of lesions together with the overwhelming presence of normal bases or nucleosides is resolved by chromatography. The targeted damage is then quantified by use of specific detection systems.

Enzymatic digestion followed by HPLC with electrochemical detection is a widely used assay for the quantification of 8-oxodGuo **44** [109, 110]. This reliable technique however can only be extended to a few other lesions, including 8-oxodAdo **48** [111], 5-OHdCyd **22** and 5-OHdUrd **23** [57], as well as the corresponding bases [112]. A more versatile technique combines gas chromatography separation with mass spectrometry detection [113, 114]. However, this assay suffers from several drawbacks [115], particularly hot acidic hydrolysis of DNA that may degrade sensitive lesions such as purine formamidopyrimidine derivatives [116, 117]. In addition, caution must be taken against artifactual oxidation that occurs during the silylation step, which is required in the preparation of volatile derivatives of DNA bases [118, 119].

More recently, high performance liquid chromatography with tandem mass spectrometry has emerged as a powerful tool to investigate DNA base damage. In this technique, the modified nucleosides are separated from normal DNA components by HPLC and the eluent is introduced into a triple quadrupolar mass spectrometer. The protonated or deprotonated pseudo-molecular ion of the targeted compound is ionized in an electrospray source and isolated by the first quadrupole. The ion is then directed towards the second quadrupole that contains a low pressure inert gas (nitrogen or argon). Collision of the accelerated pseudo-molecular ions with gas molecules leads to fragmentation of the former species. The resulting ions, whose distribution is unique to each compound, are then isolated in the third quadrupole. This tandem mass spectrometer is used as a highly specific detector in which the major fragmentation products are detected and quantified. A high sensitivity is also achieved because of low background noise. Amounts as low as a few fmol of injected material on the HPLC column can be detected. It may be added that mass spectrometry detection is quantitatively a very precise method because heavy isotopes (^{13}C,^{15}N) of the analyte may be added as internal standards in order to correct for losses of the analyte during sample preparation.

HPLC-MS/MS has been applied to the quantification of several types of radical-induced base damage within DNA [16, 17, 120–122]. In particular, five oxidized nucleosides were simultaneously detected from enzymatically digested DNA. For DNA digestion, incubation with a combination of exo- and endonucleases and phosphatase was necessary to quantitatively release all targeted compounds [17]. Formamidopyrimidine derivatives of guanine and adenine are quantified as free bases following mild room temperature formic acid hydrolysis. Quantification of oxidized nucleosides and Fapy derivatives is extremely precise because of isotopic dilution methodology.

The HPLC-MS/MS assay was also successfully applied to the measurement of UV-induced dimeric pyrimidine photoproducts [123, 124]. The latter lesions were released from DNA as modified dinucleoside monophosphates due to resistance of the intra-dimer phosphodiester group to the exonuclease activity during the hydrolysis step [125, 126]. The hydrolyzed photoproducts exhibit mass spectrometry and chromatographic features that allow simultaneous quantification of the three main classes of photolesions, namely cyclobutane dimers, (6–4) photoproducts, and Dewar valence isomers, for each of the four possible bipyrimidine sequences. It may be added that these analyses are coupled to UV detection of normal nucleosides in order to correct for the amount of DNA in the sample and obtain a precise ratio of oxidized bases or dimeric photoproducts to normal nucleosides.

3.2
Intensity Dependence on the Product Distribution

A series of experiments investigated the effect of laser pulse intensity on the distribution of damage. For each pulse intensity, DNA samples were exposed to three different doses. The quantum yield for the formation of lesions, expressed with respect to total DNA bases, was then calculated by linear regression analyses. At all intensities, the formation of lesions was found to be linear with respect to the applied dose. Oxidized nucleosides, including

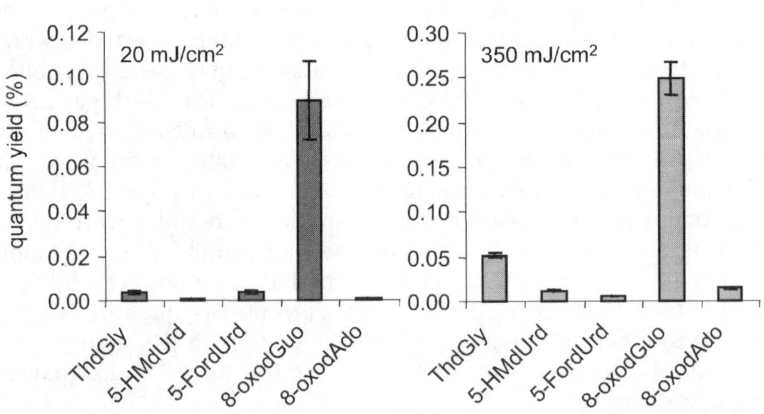

Fig. 2 Effect of the laser intensity on the quantum yield of formation of oxidized bases (ϕ_{ox}) and pyrimidine dimeric photoproducts (ϕ_{dim})

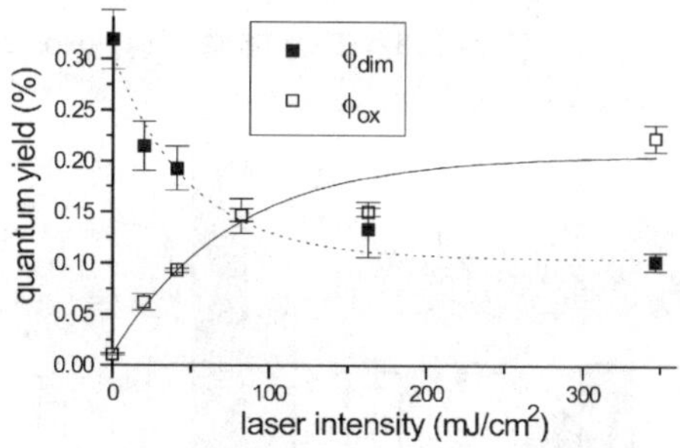

Fig. 3 Effect of UV laser intensity on the distribution of oxidation products. DNA was solubilized in 1 mM pH 7 TRIS buffer in order to stabilize the duplex

8-oxodGuo **44**, 8-oxodAdo **48**, 5-FordUrd **13**, 5-HMdUrd **12** and thymidine glycols (ThdGly) **7**, were quantified by HPLC-MS/MS, together with cyclobutane pyrimidine dimers and pyrimidine (6–4) photoproducts that arise from dimerization of adjacent pyrimidine bases via monophotonic excitation. The occurrence of DNA ionization mediated by biphotonic processes is shown by an inverse association between the quantum yield for the formation of oxidation products compared to that for the formation of dimeric pyrimidine photoproducts as a function of laser intensity (see Fig. 2). This is in agreement with previous results showing an increase in the quantum yield of Fpg-sensitive site in oligonucleotides exposed to laser pulses with increasing intensity [64].

Interestingly, 8-oxodGuo **44** was the main product among quantified oxidized bases (see Fig. 3). However, its relative yield with respect to the other oxidative lesions was found to decrease with increasing laser intensity (91 and 75% of the quantified oxidized bases at 20 and 350 mJ/cm^2, respectively, upon irradiation in 1 mM TRIS buffer). This suggests that the equilibrium between the number of guanine bases in the triplet-excited state on the one hand, and radical cation forms on the other hand, is reached at relatively low intensities in comparison to other DNA bases. This result cannot be explained by favored population of the triplet state of guanine, which is relatively high in energy compared to that of the other bases [30]. The most likely explanation involves the transfer of positive charges from other base radical cations toward guanines. Hence, the difference in the saturation intensity of the quantum yield for the formation of oxidative lesions constitutes an indirect measure of charge migration distances between bases.

4
Effect of Duplex Stability on Charge Transfer

The quantum yield for the formation of 8-oxodGuo **44** from guanine in relation to the formation of products arising from the oxidation of other nucleobases may be considered as a marker of charge transfer efficiency. This parameter was therefore used to study the influence of duplex stability on

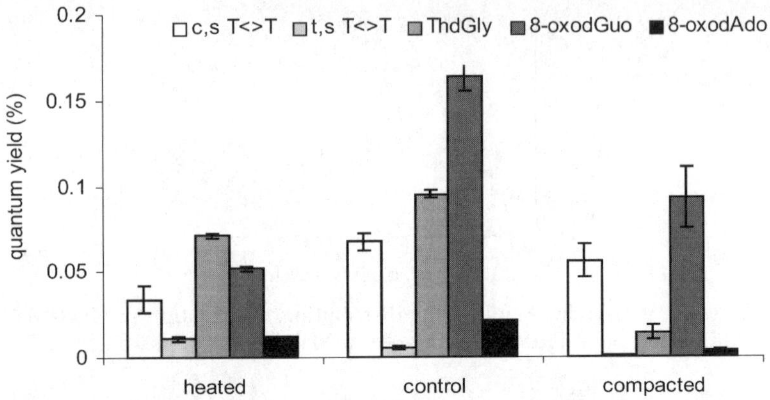

Fig. 4 Effect of DNA duplex stability on the formation of DNA lesions. Samples (75 μg/ml of calf thymus DNA) were solubilized in pure water. The different conditions were: "heated" (the solution was heated at 100°C prior to irradiation); "control" (irradiations were performed at room temperature); "compacted" (irradiations were performed in the presence of 0.1 mM spermidine)

charge transfer. The stability of DNA was controlled by modifying the ionic strength of irradiated solutions, heating the solution prior to irradiation, or adding spermidine that leads to more compacted DNA. The ratio between the yield of *cis,syn* and *trans,syn* diastereomers of thymine cyclobutane dimers (*c,s* T<>T and *t,s* T<>T, respectively) was taken as a measure of the stability of DNA duplexes. Indeed, the latter *t,s* cyclobutadithymine is hardly produced in stable B-form DNA [124]. In a first series of experiments, calf thymus DNA was diluted in pure water and heated with or without spermidine prior to irradiation. Irradiations were performed at the highest laser pulse intensity in order to favor biphotonic ionization. Thereby, it became clear that, as the stability of duplexes increased from heated to compacted DNA, the ratio between the *cis,syn* and *trans,syn* thymine cyclobutane dimers increased as well (see Fig. 4). Simultaneously, the proportion of 8-oxodGuo **44** within oxidized nucleosides also increased, reaching 32, 52 and 77% of the quantified oxidized bases in heated, control and compacted DNA, respectively. This strongly suggests that the efficiency of charge transfer increases upon stabilization of DNA duplexes.

The role of duplex stability on charge transfer was confirmed by the effect of laser intensity on the relative yield of 8-oxodGuo **44**. Indeed, the results indicate that charge transfer is directed towards guanine residues because the quantum yield of formation of 8-oxodGuo **44** reaches a point of saturation at much weaker intensity compared to that of other modifications. Therefore, one might anticipate that the difference in the relative yield of 8-oxodGuo **44** between low and high intensity would be small under conditions of efficient charge transfer. This was actually observed in experiments in which the ratio of 8-oxodGuo **44** at low (20 mJ/m^2) to high (350 mJ/m^2) intensities was 2.1, 1.5 and 1.1 for heated, control and compacted DNA. Similar observations, in terms of the relative yield of 8-oxodGuo **44** at high intensity and the effect of pulse intensity on the distribution of oxidative damage, were made when the DNA duplex was stabilized by the addition of increasingly higher concentrations of TRIS.

The observation that stabilization of the DNA duplex favors charge transfer sheds some light on the mechanism of this process. Two very different mechanisms of charge transfer have been advanced, which include long distance coupling between bases within a π-stack [127, 128] and hopping of the electron hole between sites of low ionization potential (guanine or GG doublets) [129, 130]. A more refined version of the latter process involves delocalization of the positive hole between several bases similar to a polaron [11, 131]. Interestingly, theoretical calculations based on the latter model led to the conclusion that charge transfer efficiency increases with increasing duplex stability [132]. This contrasts with other calculations based on the hypothesis of super-exchange that predicts an increase in charge transfer with DNA fluctuations [133, 134]. On the basis of results with biphotonic ionization of DNA and HPLC-MS/MS analysis of the photoproducts, the former hypothesis seems to be the most important.

5
Conclusion

Evidence is provided in this chapter that HPLC-MS/MS quantification of specific one-electron oxidation products of purine and pyrimidine bases is a powerful tool for the study of charge transfer processes within DNA. Although this approach does not give sequence information about the damage, it does give unambiguous identification and quantitation of several DNA modifications arising from one-electron oxidation. This approach is complementary to the common assay that involves mapping DNA lesions at nucleotide resolution after cleavage of defined sequence defined DNA fragments at the site of the damage by either hot piperidine treatment or upon incubation with DNA *N*-glycosylase repair enzymes. Indeed, the latter assay does not provide unambiguous identification of the modified bases.

It should be added that some lesions are resistant to the chemical or enzymatic treatments aimed at inducing strand breaks, and so they are not revealed, leading to loss of structural and mechanistic information. Another point of concern is the occurrence of secondary oxidation reactions that involve 8-oxodGuo **44**, a major product of guanine oxidation that is susceptible to further oxidation (*vide supra*). Therefore, it will be necessary to increase the sensitivity of detection in the methods used for mapping oxidative base damage at the sequence level in order to detect damage at a low overall extent, and minimize the occurrence of secondary reactions. Additional efforts should be made to allow for the detection of damage at the nucleotide level in a more specific and also more extensive manner, including for example the measurement of FapyGua **43** and oxidation products of adenine. To achieve this goal, we will need to carefully delineate the stability of lesions to alkali treatment and of the substrate specificity of DNA glycosylases used to reveal the presence of chemical damage within sequence defined DNA probes. The design of specific oxidation conditions that convert base damage such as 8-oxodAdo **48** into piperidine labile sites or that are recognized by DNA repair *N*-glycosylases should be researched.

References

1. Sies H (1991) Oxidative Stress, Oxidants and Antioxidants. Academic Press, New York
2. Lindahl T (1993) Nature 362:709
3. Ames BN, Gold LS, Willett WC (1995) Proc Natl Acad Sci USA 92:5258
4. Loft S, Poulsen HE (1996) J Mol Med 74:297
5. Marnett LJ, Burcham PC (1993) Chem Res Toxicol 6:771
6. Cadet J (1994) In: Hemminki K, Dipple A, Shuker DEG, Kadlubar FF, Sagerbäck D, Barstch D (eds) DNA Adducts: Identification and Biological Significance. International Agency for Research on Cancer. IARC Scientific Publications, Lyon No 125: p 245
7. Cadet J, Berger M, Douki T, Ravanat JL (1997) Rev Physiol Biochem Pharmacol 131:1
8. Burrows CJ, Muller JG (1998) Chem Rev 98:1109
9. Cadet J, Delatour T, Douki T, Gasparutto D, Pouget JP, Ravanat JL, Sauvaigo S (1999) Mutat Res 424: 9
10. Cai Z, Sevilla MD (2003) In: G. Schuster (ed) Long-range electron transfer efficiency within DNA, Topics in Current Chemistry. Springer, Berlin Heidelberg New York (this volume)
11. Henderson PT, Jones D, Hampikian G, Kan Y, Schuster GB (1999) Proc Natl Acad Sci USA 96:8353
12. Giese B, Wesselky S, Sportmann M, Lindemann U, Meggers E, Michel-Beyerle ME (1999) Angew Chem Int Ed 38:996
13. Giese B, Spichty M (2000) Chem Phys Chem 1:195
14. Wan C, Fiebig T, Schiermann O, Barton JK, Zewail AH (2000) Proc Natl Acad Sci USA 97:14052
15. Giese B, Amaudrut J, Köhler A-K, Spormann M. Wessely S. (2001) Nature 412:318
16. Ravanat JL, Duretz B, Guiller A, Douki T, Cadet J (1998) J Chromatogr B Biomed Sci Appl 715:349
17. Frelon S, Douki T, Ravanat JL, Pouget JP, Tornabene C, Cadet J (2000) Chem Res Toxicol 13:1002
18. Cadet J, Douki T, Frelon S, Sauvaigo S, Pouget J-P, Ravanat J-L (2002) Free Radic Biol Med 33:441
19. Becker D, Sevilla M (1993) Adv Radiat Biol 17:121
20. Symons MCR (1995) Radiat Phys Chem 45:837
21. Debije MG, Milano MT, Bernhard WA (1999) Angew Chem Int Ed 38:2752
22. Cai Z, Li X, Sevilla M D (2002) J Phys Chem B 106:2755
23. Swarts SG, Becker D, Sevilla M, Wheeler KT (1996) Radiat Res 145:304
24. Shaw AA, Cadet J (1990) J Chem Soc Perkin Trans 2:2063
25. Gromova M, Balanzat E, Gervais B, Nardin R, Cadet J (1998) Int J Radiat Biol 74:81
26. Gromova M, Nardin R, Cadet J (1998) J Chem Soc Perkin Trans 2:1365
27. Fisher GJ, Land EJ (1983) Photochem Photobiol 37:27
28. Wagner JR, van Lier JE, Johnston LJ (1990) Photochem Photobiol 52:333
29. Cadet J, Vigny P (1990) In: Morrison H (ed) Bioorganic Photochemistry. Photochemistry and the Nucleic Acids. Wiley and Sons, New York, vol 1, p 1
30. Nikogosyan DN (1990) Int J Radiat Biol 57:233
31. Candeias LP, Stennken S (1992) J Am Chem Soc 114:699
32. Wagner JR, van Lier JE, Decarroz C, Berger M, Cadet J (1990) Methods Enzymol 186:502
33. Wagner JR, van Lier JE, Berger M, Cadet J (1994) J Am Chem Soc 116:2235

34. von Sonntag C (1987) In: The Chemical Basis of Radiation Biology. Taylor Francis, London
35. Decarroz C, Wagner JR, van Lier JE, Krishna CM, Riesz P, Cadet J (1986) Int J Radiat Biol 50:491
36. Krishna CM, Decarroz C, Wagner JR, Cadet J, Riesz P (1987) Photochem Photobiol 46:175
37. Isildar M, Schuchmann MN, Schulte-Frohlinde D, von Sonntag C (1982) Int J Radiat Biol 41:525
38. Fujita S, Steenken S (1981) J Am Chem Soc 103:2540
39. Jolibois F, D'Ham C, Grand A, Subra R, Cadet J (1998) J Mol Struct (Theochem) 427:143
40. Wagner JR, Berger M, Cadet J, van Lier JE (1990) J Chromatogr 504:191
41. Cadet J, Ulrich J, Téoule R (1975) Tetrahedron 31:2057
42. Cadet J, Ducolomb R, Hruska FE (1979) Biochim Biophys Acta 563:206
43. Cadet J, Nardin R, Voituriez L, Remin M, Hruska FE (1981) Can J Chem 59:3313
44. Hruska FE, Sebastian R, Grand A, Voituriez L, Cadet J (1987) Can J Chem 65:2618
45. Lutsig MJ, Cadet J, Boorstein RJ, Teebor GW (1992) Nucleic Acids Res 20:4839
46. Bardet M, Cadet J, Wagner RJ (1996) Magn Res Chem 34:577
47. Maufrais C, Fazakerley GV, Cadet J, Boulard Y (2000) Biochemistry 39:5614
48. Douki T, Delatour T, Paganon F, Cadet J (1996) Chem Res Toxicol 9:1145
49. Bjelland S, Birkeland NK, Benneche T, Volden G, Seeberg E (1994) J Biol Chem 269:30489
50. Bjelland S, Eide L, Time RW, Stote R, Eftedal I, Volden G, Seeberg E (1995) Biochemistry 34:14758
51. Douki T, Cadet J (1999) Int J Radiat Biol 75:571
52. Saito I, Takayama M, Kawanishi S (1995) J Am Chem Soc 117: 5590
53. Wagner JR, Decarroz C, Berger M, Cadet J (1999) J Am Chem Soc 121:4101
54. Tremblay S, Douki T, Cadet J, Wagner JR (1999) J Biol Chem 274:20833
55. Decarroz C, Wagner JR, Cadet J (1987) Free Rad Res Commun 2:295
56. Roupioz Y, Lhomme J, Kotera M (2002) J Am Chem Soc 124:9129
57. Wagner JR, Hu CC, Ames BN (1992) Proc Natl Acad Sci USA 89:3380
58. Bienvenu C, Wagner JR, Cadet J (1996) J Am Chem Soc 118:11406
59. Russell GA (1957) J Am Chem Soc 79:3871
60. Simic M, Hayon E (1974) FEBS Lett 44:334
61. Bienvenu C, Lebrun C, Cadet J (1997) J Chim Phys 94:300
62. Bienvenu C, Cadet J (1996) J Org Chem 61:2632
63. Yokoya A, Cunniffe SM, O'Neill P (2002) J Am Chem Soc 124:8859
64. Angelov D, Spassky A, Berger M, Cadet J (1997) J Am Chem Soc 119:11373
65. Douki T, Angelov D, Cadet J (2001) J Am Chem Soc 123:11360
66. O'Neill P, Parker AW, Plumb MA, Siebbeles LDA (2001) J Phys Chem B 105:5283
67. Ito K, Kawanishi S (1997) Biochemistry 36:1774
68. Saito I, Nakamura T, Nakatani K (2000) J Am Chem Soc 122:3001
69. Luxford C, Dean RT, Davies MJ (2000) Chem Res Toxicol 13:665
70. Shafirovich V, Dourandin A, Huang W, Geacintov NE (2001) J Biol Chem 276:24621
71. Milligan JR, Aguilera JA, Nguyen JV, Ward JF (2001) Int J Radiat Biol 77:281
72. Steenken S, Jovanovic SV (1997) J Am Chem Soc 119:617
73. Cadet J, Berger M, Buchko GW, Joshi P, Raoul S, Ravanat J-L (1994) J Am Chem Soc 116:7403
74. Raoul S, Berger M, Buchko GW, Joshi PC, Morin B, Weinfeld M, Cadet J (1996) J Chem Soc Perkin Trans 2:371
75. Candeias LP, Steenken S (2000) Chem Eur J 6:475
76. Vialas C, Pratviel G, Claparols C, Meunier B (1998) J Am Chem Soc 120:11548
77. Gasparutto D, Ravanat J-L, Gérot O, Cadet J (1998) J Am Chem Soc 120:10283
78. Kino K, Saito I, Sugyiama H (1998) J Am Chem Soc 120:7373
79. Cullis PM, Malone ME, Merson-Davies LA (1996) J Am Chem Soc 118:2775
80. Morin B, Cadet J (1995) Chem Res Toxicol 8:792

81. Morin B, Cadet J (1995) J Am Chem Soc 117:12408
82. Steenken S (1989) Chem Rev 89:503
83. Douki T, Spinelli S, Ravanat J-L, Cadet J (1999) J Chem Soc Perkin Trans 2:1875
84. Kasai H, Yamaizumi Z, Berger M, Cadet J (1992) J Am Chem Soc 114:9692
85. Haraguchi K, Greenberg MM (2001) J Am Chem Soc 123:8636
86. Greenberg MM, Hantosi Z, Wiederholt CJ, Rithner CD (2001) Biochemistry 40:15856
87. Sambandam CJ, Hantosi Z, Greenberg MM (2002) J Am Chem Soc 124:3263
88. Burgdorf LT, Carell T (2002) Chemistry 8:293
89. Wagenknecht H-A, Rajski SR, Pascaly M, Stemp EDA, Barton JK (2001) J Am Chem Soc 123:4400
90. Raoul S, Bardet M, Cadet J (1995) Chem Res Toxicol 8:924
91. Frelon S, Douki T, Cadet J (2002) Free Radic Res 36:499
92. Wang Y, Liu Z, Dixon C (2002) Biochem Biophys Res Commun 291:1252
93. Bernstein R, Prat F, Foote CS (1999) J Am Chem Soc 121:464
94. Duarte V, Muller JG, Burrows CJ (1999) Nucleic Acids Res 27:496
95. Luo W, Muller JG, Rachlin EM, Burrows CJ (2001) Chem Res Toxicol 14:927
96. Luo W, Muller JG, Rachlin EM, Burrows CJ (2000) Org Lett 2:613
97. Sugden KD, Campo CK, Martin BD (2001) Chem Res Toxicol 14:1315
98. Adam W, Kurz A, Saha-Möller CR (2000) Chem Res Toxicol 13:1199
99. Adam W, Arnold MA, Grüne M, Nau WM, Pischel U, Saha-Möller CR (2002) Org Lett 4:537
100. Doddridge AZ, Cullis PM, Jones GDD, Malone ME (1998) J Am Chem Soc 120:10998
101. Luo W, Muller JG, Burrows CJ (2001) Org Lett 3:2801
102. Steenken S, Jovanovic, SV, Bietti M, Bernhard K (2000) J Am Chem Soc 122:2373
103. Niles JC, Wishnok JS, Tannenbaum SR (2001) Org Lett 3:963
104. Tretyakova NY, Wishnok JS, Tannenbaum SR (2000) Chem Res Toxicol 13:658
105. Suzuki T, Masuda M, Friesen MD, Ohshima H (2001) Chem Res Toxicol 14:1163
106. Hickerson RP, Prat F, Muller JG, Foote CS, Burrows CJ (1999) J Am Chem Soc 121:9423
107. Taylor J-S (1990) J Chem Educ 67:835
108. Douki T, Cadet J (1995) In: Jollès P, Jornvall H (eds) Birkhäuser, Basel, p 173
109. Floyd RA, Watson JJ, Wong PK, Altmiller DH, Rickard RC (1986) Free Rad Res Commun 1:163
110. Kasai H (1997) Mutat Res 387:147
111. Berger M, Anselmino C, Mouret J-F, Cadet J (1990) J Liq Chromatogr 13:932
112. Kaur H, Halliwell B (1996) Biochem J 318:21
113. Dizdaroglu M, Bergtold DS (1986) Anal Biochem 156:182
114. Dizdaroglu M Gajewski E (1990) Methods Enzymol 186:530–544
115. Cadet J, D'Ham C, Douki T, Pouget J-P, Ravanat J-L, Sauvaigo S (1998) Free Radic Res 29:541
116. Douki T, Martini R, Ravanat J-L, Turesky RJ, Cadet J (1997) Carcinogenesis 18:2385
117. Swarts SG, Smith GS, Miao L, Wheeler KT (1996) Radiat Environ Biophys 35:41
118. Douki T, Delatour T, Bianchini F, Cadet J (1996) Carcinogenesis 17:347
119. Ravanat J-L, Turesky RJ, Gremaud E, Trudel LJ, Stadler RH (1995) Chem Res Toxicol 8:1039
120. Serrano J, Palmeira CM, Wallace KB, Kuehl DW (1996) Rapid Commun Mass Spectrom 10:1789
121. Ravanat J-L, Remaud G, Cadet J (2000) Arch Biochem Biophys 374:118
122. Frelon S, Douki T, Cadet J (2002) Free Radic Res 36:499
123. Douki T, Court M, Sauvaigo S, Odin F, Cadet J (2000) J Biol Chem 275:11678
124. Douki T Cadet J (2001) Biochemistry 40:2495
125. Liuzzi M, Weinfeld M, Paterson MC (1989) J Biol Chem 264:6355
126. Douki T, Zalizniak T, Cadet, J (1997) Photochem Photobiol 66:171
127. Hall DB, Holmlin RE, Barton JK (1996) Nature 382:731
128. Holmlin RE, Dandliker PJ, Barton JK (1998) Angew Chem Int Ed 36:2715
129. Giese B (2000) Acc Chem Res 33:631

130. Bixon M, Giese B, Wessely S, Langenbacher T, Michel-Beyerle ME, Jortner J (1999) Proc Natl Acad Sci USA 96:11713
131. Schuster GB (2000) Acc Chem Res 33:253
132. Conwell EM, Rakhmanova SV (2000) Proc Natl Acad Sci USA 97:4556
133. Bruinsma R, Grüner G, D'Orsogna MR, Rudnick J (2000) Phys Rev Lett 85:4393
134. Troisi A, Orlandi G (2002) J Phys Chem B 106:2093

Top Curr Chem (2004) 236:27–44
DOI 10.1007/b94410

Hole Injection and Hole Transfer Through DNA: The Hopping Mechanism

Bernd Giese

Department of Chemistry, University of Basel, St Johanns Ring 19, 4056 Basel, Switzerland
E-mail: bernd.giese@unibas.ch

Abstract Hole injection into a guanine of DNA, and hole transfer between the DNA guanines were studied. The reactions were performed in the ground state using a carbohydrate radical cation whose precursor is site-selectively incorporated into the DNA. For the charge injection step the influence of the distance, the energy difference of donor and acceptor, and the bridge were measured. The guanine radical cation, generated in the injection step, triggered long distance charge transfer through DNA, as measured by the water trapping products. It turned out that the guanines are the charge carriers and, if the distance between the guanines is long, adenines also carry the positive charge. Therefore, the positive charge migrates through the DNA over long distances in a multistep hopping process. Trapping of the positive charge at the DNA bases either by nucleophilic attack or by proton transfer to the surrounding water stops the hole transfer through DNA. This can be induced, for example, by mismatches of the G:C base pairs.

Keywords Charge injection · Charge transfer · Hopping mechanism · Mismatches

1 Introduction

Assays developed for the study of long distance charge transfer through DNA have to include the following experimental steps:

1. Attachment of the charge injection system to DNA
2. Injection of the charge into a DNA base
3. Migration of the charge through DNA
4. Detection of the charge

As described in this book, the individual research groups use different charge injection systems, which contain artificial molecules attached in different ways to the DNA. Their activation by light, reduction, or oxidation leads to systems with different redox potentials. Therefore, the observed rates of charge injection into a DNA base vary from assay to assay. Once a DNA base is oxidized or reduced, it triggers charge transfer through DNA, which should be mainly independent of the injection system. But the method of charge detection in the DNA strand can influence the experimental results. For example, a chemical trapping of the charge depends upon the reagent and the reaction conditions. Therefore, a general picture of the charge transfer through DNA can be concluded from experimental results only if the different experimental conditions for the charge injection and the charge detection are well understood and taken into account. A simple, direct comparison of the experimental results that neglects the boundary conditions of the various assays leads to confusion. A similar warning is given by K. Nakatani and I. Saito in their article of this volume.

We will discuss an electron hole injection system that is quite different from all the other ones described in this book.

2 Hole Injection

2.1 Injection System

Our method of hole injection into DNA is based on the higher oxidation potential of the enol ether radical cation **1** compared to the guanine radical cation **4**. The generation of an enol ether radical cation **7**, which can be used for hole transfer experiments in DNA, starts from the modified nucleotide **5** [1], whose photolysis forms the nucleotide radical **6**. In double and single stranded oligomers radical **6** cleaves the β-C,O-bond with reaction rates of 10^2–$10^3 s^{-1}$ [2] leading to radical cation **7** in high yields [3] as long as radical traps like O_2 are absent (Scheme 1).

Enolether radical cation **7**, generated in a double strand, oxidizes selectively a nearby guanine base (G) in the same double strand [3, 4]. This elec-

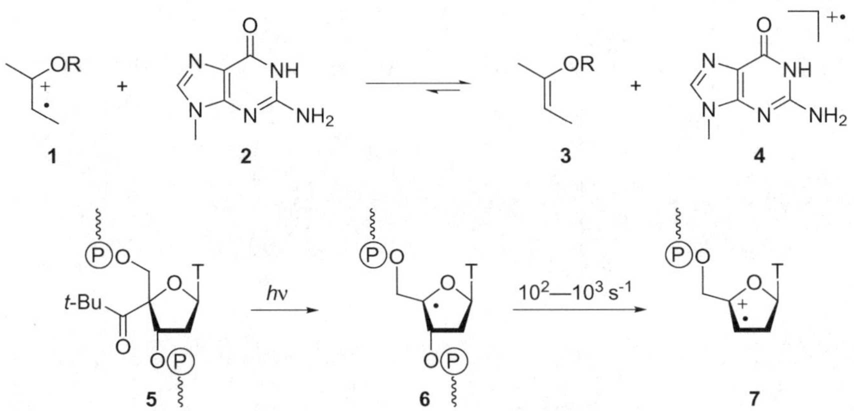

Scheme 1 Generation of guanine radical cation **4** and sugar radical cation **7**

tron transfer step (**7**→**8**) competes with the addition of water to the radical cation **7**, which leads to products **9** and **10** in irreversible, pseudo first order reactions (Scheme 2). The three products **8**, **9** and **10** can be separated and

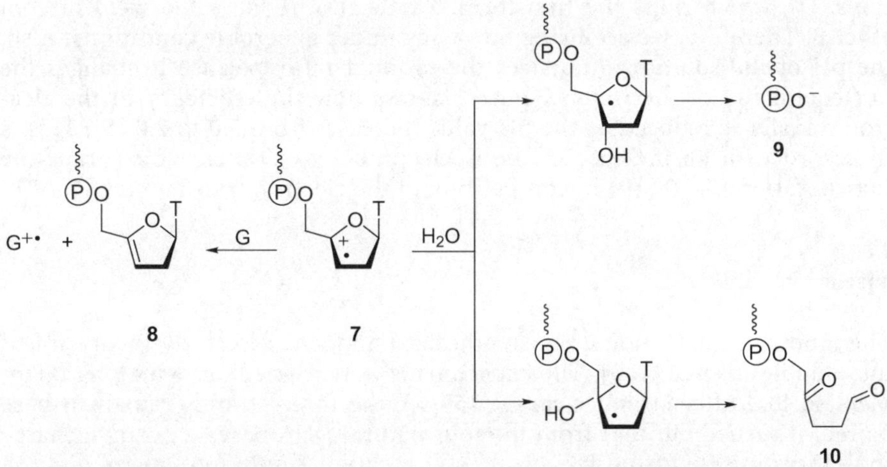

Scheme 2 Competition between electron transfer (**7**→**8**) and water trapping (**7**→**9**+**10**) of sugar radical cation **7**

quantified by HPLC [3]. They are the only products formed from the radical cation **7** under our conditions. In test experiments, we have shown that the water trapping reaction of **7** does not depend upon the DNA strand. Therefore the product ratios **8**/(**9**+**10**) are the relative rates of the charge injection step into a guanine of the DNA. Because we have determined the pseudo first

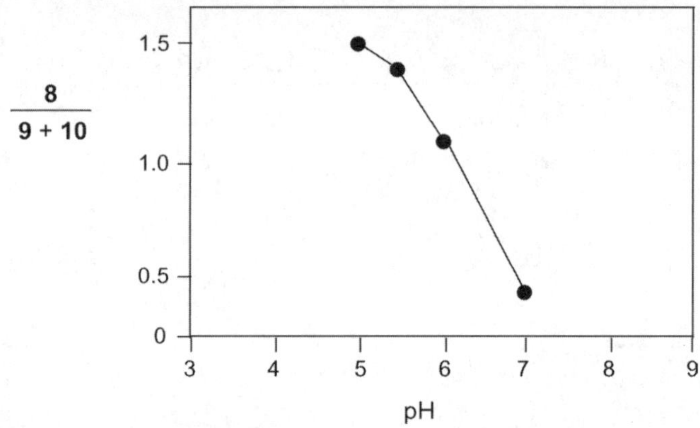

Fig. 1 Influence of the pH value on the ratio of the electron transfer, forming enol ether **8**, and the water trapping, yielding products **9+10**, of the enol radical cation **7** in a DNA double strand

order rate of the water trapping to $10^8 s^{-1}$ (pH 5.0), the absolute electron transfer rates could be measured in competition experiments [3].

The efficiency of this injection system depends upon the reaction conditions. O_2, which traps the first formed radical **6**, reduces the yield of enol ether **8**. Therefore, we are using our assay under anaerobic conditions. Also, the pH of the solution influences the product ratio because it changes the nucleophilicity of the water. Figure 1 shows how the efficiency of the electron transfer is reduced as the pH value increases from 5.0 to 7.0 [5]. This is in accord with an increase of the nucleophilicity of water, which traps the radical cation (**7**→**9+10**) in competition to the electron transfer step (**7**→**8**).

2.2
Injection into DNA

The modified nucleotide **5** was synthesized and site selectively incorporated into double strands [1, 6]. The experiments were carried out with long oligomers so that after strand cleavage (**5**→**7**), the three strands remained base paired. It turned out that from the four natural DNA bases a positive charge (hole) was injected from the sugar radical cation **7** only into guanine [4, 7], the base of lowest oxidation potential [8]. With A:T base pairs as possible electron donors the water trapping of the enol ether radical cation (**7**→**9+10**) was faster than a charge injection (**7**→**8**). We have studied the details of the charge injection step, where a G:C base pair is the electron donor, the $(A{:}T)_n$ sequence is the bridge, and the enol ether radical cation is the electron acceptor (Scheme 3).

According to the Marcus theory [9], the electron transfer rate depends upon the reaction enthalpy (ΔG), the electronic coupling (V) and the reorganization energy (λ). By changing the electron donor and the bridge we measured the influence of these parameters on the charge transfer rate. The re-

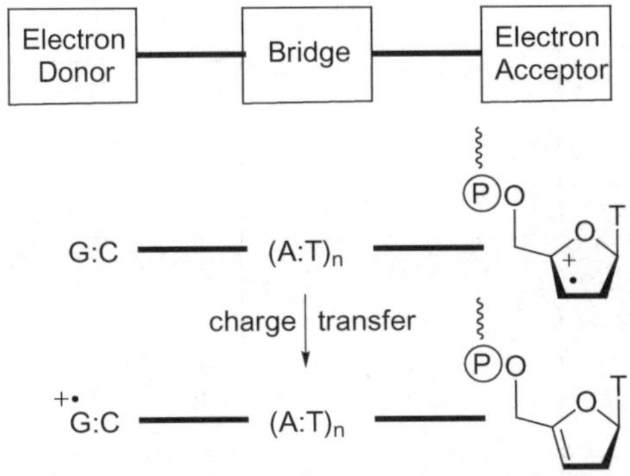

Scheme 3 Charge transfer from the sugar radical cation to guanine (G)

action enthalpy was varied by exchanging guanine (G) by 8-oxoguanine (G^{oxo}) and 7-deazaguanine (G^{Z}). The data in Fig. 2 show that the decrease of the ionization energy of these electron donors [10] leads to an increase of

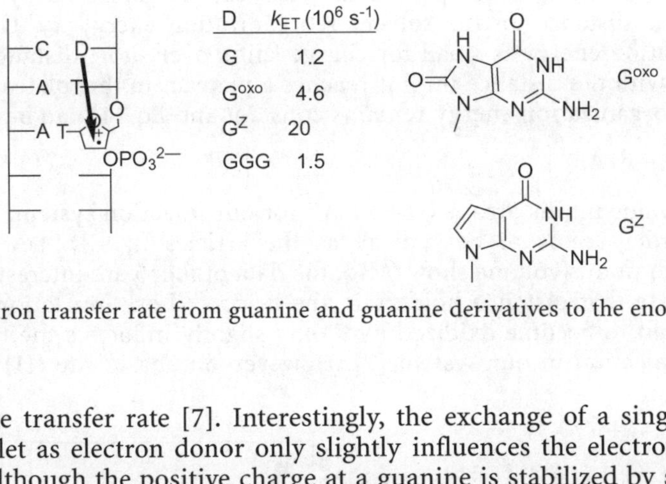

Fig. 2 Electron transfer rate from guanine and guanine derivatives to the enol ether radical cation

the charge transfer rate [7]. Interestingly, the exchange of a single G by a GGG triplet as electron donor only slightly influences the electron transfer rate [4] although the positive charge at a guanine is stabilized by surrounding guanines. Similar results were also obtained by time resolved experiments of M.E. Michel-Beyerle [11] as well as F.D. Lewis (see his article in this volume).

The influence of the electronic coupling on the electron transfer rate was determined by changing the length of the $(A{:}T)_n$ bridge. As expected, the rate decreased as the number n of the A:T base pairs between electron donor and electron acceptor increased [4, 7]. But, surprisingly, the exponential correlation of Eq. (1) between the rate k_{ET} and the distance is not valid for short distances. The plots in Fig. 3 and Fig. 4 show that at 6 Å the electron transfer rate k_{ET} is much faster than expected [4, 7].

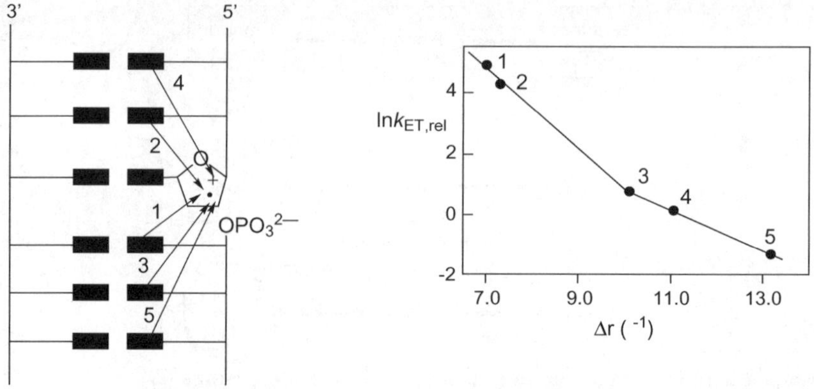

Fig. 3 Distance dependence of the relative electron transfer rate from a single guanine to the enol ether radical cation

Such a rate increase at short distances has been observed also by M.E. Michel-Beyerle [12] in time resolved experiments with a photoactivated acridinium ion as electron acceptor. This effect can be explained by the influence of the distance on the solvent reorganization energy λ_S. The solvent reorganization energy is small for charge shifts over short distances, and it increases with the distance until it reaches a plateau. In this plateau area the solvent reorganization energy remains constant and Eq. (1) can be applied:

$$\ln k_{ET} \propto -\beta \cdot \Delta r \qquad (1)$$

The β-value in this area is 0.6±0.1 Å^{-1} for our injection system. A β-value in this order seems to be typical, as the articles by F.D. Lewis and V. Shafirovich in this volume show. Also, the data of Fig. 5 are interesting. They demonstrate that distance holders in the bridge whose ionization energies are too high to become oxidized by 7 only slightly influence the rate of the electron injection in our system [7]. However, an abasic site (H) increases

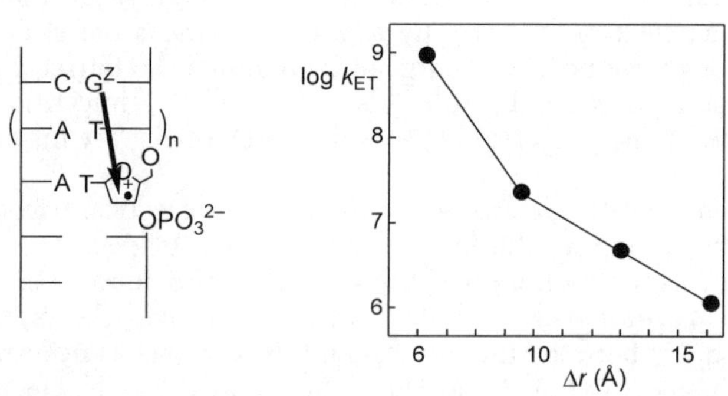

Fig. 4 Distance dependence of the electron transfer rate from 7′-deazaguanine G^Z (see Fig. 2) to the enol ether radical cation

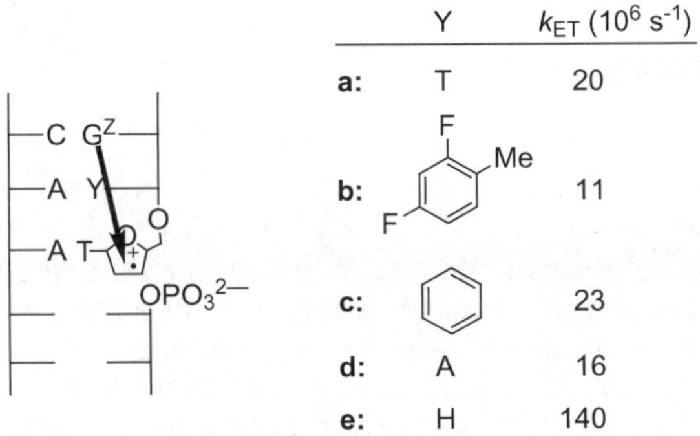

Fig. 5 Influence of the bridging substituent Y on the electron transfer rate from a 7′-deazaguanine G^Z (see Fig. 2)

the rate, presumably because it decreases the distance between donor and acceptor by enhancing the flexibility of the DNA strand [7].

2.3 Injection into Single Strands

Single stranded oligomers exhibit a completely different behavior than double strands. As shown in Fig. 6, the increase of the number *n* of thymine bases in the bridge between the electron donor guanine and the enol ether radical cation as electron acceptor influence the rate of the electron transfer only slightly. This can be explained by the flexibility of the single strand, which levels out the distance between donor and acceptor [13].

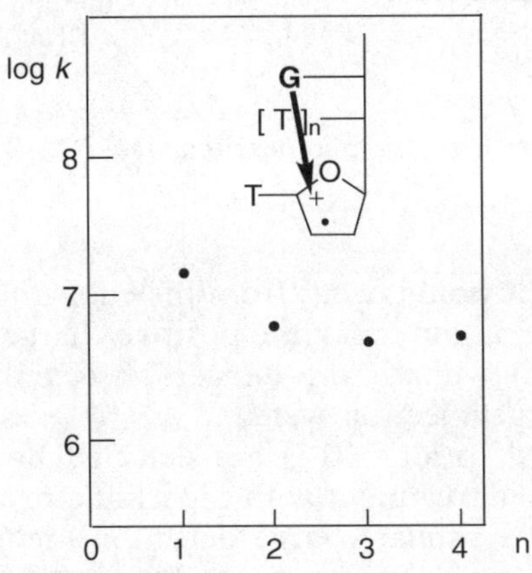

Fig. 6 Hole injection into a guanine (G) in single strands, which is separated from the enol ether radical cation by various numbers of thymidines $[T]_n$

3
Hole Transfer through DNA

3.1
Analytical Method

Having injected the positive charge into a guanine base, the question arises as to whether this guanine radical cation ($G^{\cdot+}$) can induce charge transfer through DNA. In assays like **11** where the guanine is oxidized by a photoexcited injection system (**11**→**12**), the back electron transfer (**12**→**11**) is a severe problem because it is exothermic and therefore often faster than the thermoneutral electron transfer from a guanine within the DNA duplex (**12**→**13**) (Scheme 4). In our system **14** the back electron transfer (**15**→**14**) is

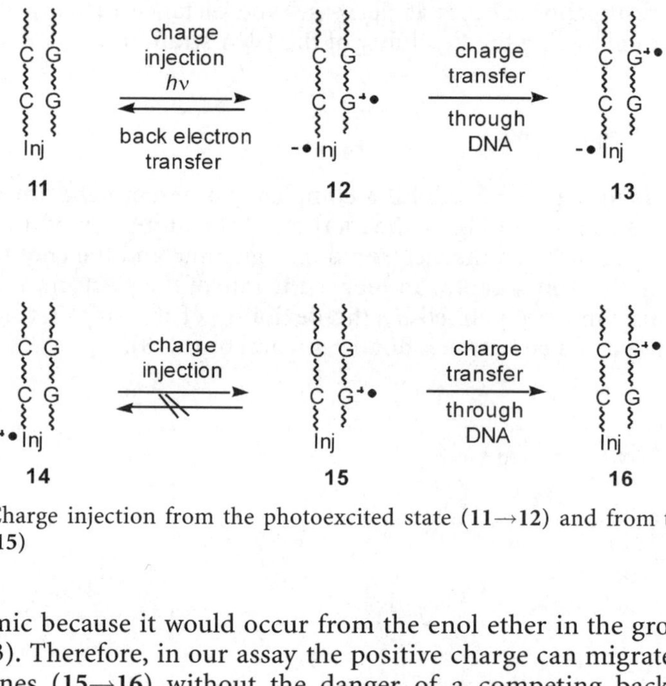

Scheme 4 Charge injection from the photoexcited state (**11**→**12**) and from the ground state (**14**→**15**)

endothermic because it would occur from the enol ether in the ground state (Scheme 3). Therefore, in our assay the positive charge can migrate between the guanines (**15**→**16**) without the danger of a competing back electron transfer from the charge injection system.

The guanine radical cations ($G^{\cdot+}$) are detected by their reactions with water, which leads after treatment with piperidine or ammonia to selective strand cleavage [14]. A similar charge detection method was used by J.K. Barton, G.B. Schuster and I. Saito as described in their articles in this volume. The cleavage products were separated and quantified by gel electrophoresis. A typical example is shown in Fig. 7 where the GGG unit acts as a thermodynamic sink for the positive charge, and the efficiency of the charge transfer can be measured by the product ratio P_{GGG}/P_G.

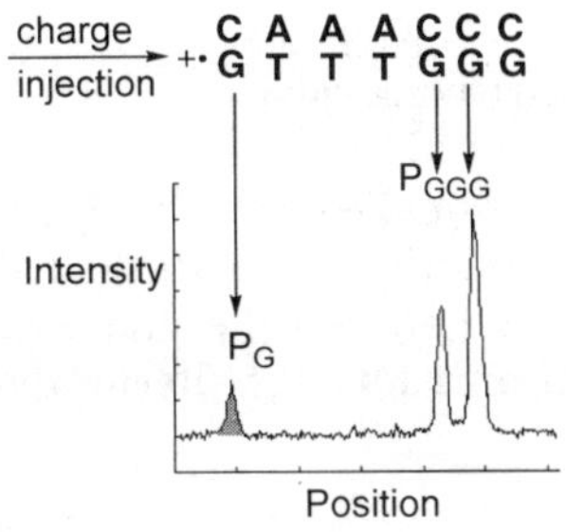

Fig. 7 Histogram showing the products P_G and P_{GGG}, formed after charge injection into G, water trapping of the guanine radical cations and subsequent strand cleavage

But the ratio of the products P_{GGG}/P_G depends not only upon the charge transfer but also upon the water trapping reaction rate [15]. This water trapping of a guanine radical cation should change if the reaction conditions are modified. For example, in strand **17** (Fig. 8), the product ratio P_{GGG}/P_{G1} decreases from 30 to 9 if the pH value is increased from 5.0 to 7.0 [16]. Obviously, the water trapping of the guanine radical cation $G_1^{\cdot+}$, which competes with the hole transfer to GGG, is more efficient at pH 7.0 than at pH 5.0, which can be explained by an increase of the water nucleophilicity (see also Fig. 1).

Also, the distance of the injection system from the first oxidized guanine influences the product ratio P_{GGG}/P_{G1}, because the injection system itself can influence the water trapping rate of the nearby $G_1^{\cdot+}$. Therefore, the experimentally detected efficiencies of the charge transfer (measured by the ratio of the chemical products) depend upon the special injection system and the reaction conditions. This has to be taken into account if the experimental results obtained under different reaction conditions are to be compared with each other.

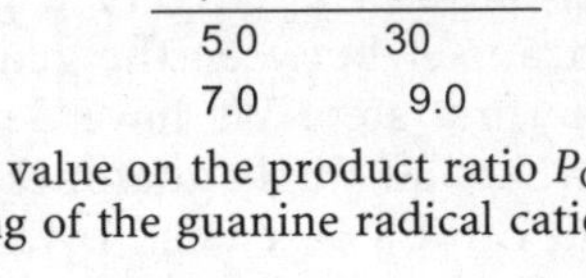

17

pH	P_{GGG}/P_G
5.0	30
7.0	9.0

Fig. 8 Influence of the pH value on the product ratio P_{GGG}/P_G, formed after charge injection into G, water trapping of the guanine radical cations and subsequent strand cleavage

3.2
Single Step Charge Transfer Between Guanines

In order to determine the influence of the bridge length on the efficiency of the hole transfer between guanines, we studied double strands where the number of $(A{:}T)_n$ base pairs between the guanines were varied (Fig. 9). In these experiments the GGG unit acts as a thermodynamic sink for the charge [17].

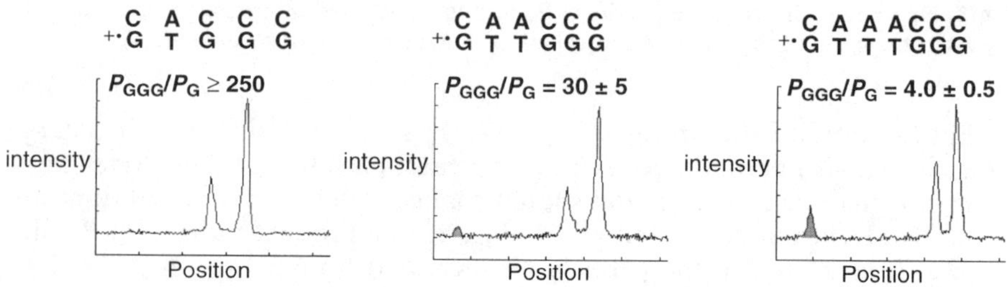

Fig. 9 Histograms for three different strands showing the products P_G and P_{GGG} formed after charge injection into G, water trapping of the guanine radical cations and subsequent strand cleavage

The data of Fig. 9 show that an increase of the $(A{:}T)_n$ sequence from n=13 drastically decreases the efficiency (P_{GGG}/P_G) of the charge transfer. A plot of these ratios against the distance Δr according to Eq. (2) gives a β-value of 0.6 ± 0.1 Å^{-1}.

$$\log P_{GGG}/P_G \propto^{\sim} -\beta \cdot \Delta r \tag{2}$$

Similar results are described by K. Nakatani and I. Saito in their article of this volume. The distance effect is in accord with a single step charge shift between the guanines where the A:T base pairs act as bridge that are not oxidized during the reaction, as described in the articles of J. Jortner, F.D. Lewis as well as D. Beratan and M. A. Ratner in this volume.

3.3
G-Hopping

In DNA sequences where the water trapping of a guanine radical cation ($G^{\cdot +}$) is slower than the hole transfer between the guanine bases, the charge shift should not stop after the first step. We have demonstrated this by an experiment shown in Fig. 10 where all single guanines lead to water trapping products [18]. The experiment can be described by a reversible diffusion of the charge between the guanine charge carriers so that the overall charge transport over long distances occurs in a multistep hopping process. The experiments of K. Nakatani and I. Saito, described in this volume, also show that single guanines are carriers of the positive charge. The theoretical treat-

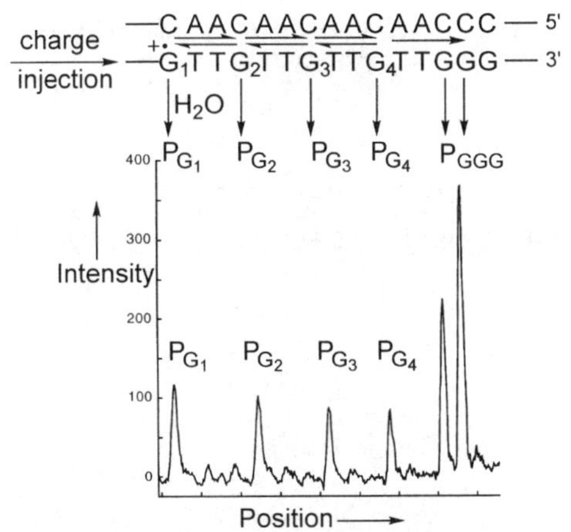

Fig. 10 Histogram showing the products P_G and P_{GGG} formed after charge injection into G_1, water trapping of the guanine radical cations and subsequent strand cleavage

ment of such a hopping reaction is described in the articles of J. Jortner as well as D. Beratan and M.A. Ratner.

But the experimental results of hopping experiments using this assay can be easily overinterpreted or misunderstood [15] because long distance hole hopping occurs only if the water trapping is so slow that the charge transfer can compete with it. Often, the rate determining steps for the formation of products P_G are the water trapping reactions, and the decrease of the products P_{G1} to P_{Gn} does not simply reflect the distance influence of the hole transfer rate (Fig. 11).

In extreme cases where all electron transfer steps are reversible and the water trapping reactions are very slow, the charge is distributed over the guanines according to the thermodynamic stabilization. From these experiments one cannot deduce the influence of the sequence on the hole transfer rate. Therefore, using a chemical assay of this type leads to results that have to be discussed with great care.

Nevertheless, long distance charge transfer through DNA is possible only because of a hopping mechanism. A single step charge shift over long distances induced by a guanine radical cation ($G^{\cdot+}$) is much too slow to compete with the trapping reaction of $G^{\cdot+}$ by water.

3.4 A-Hopping

In section 3.2. we described how the efficiency of hole transfer between a G and a GGG triplet strongly depends upon the length of the $(A{:}T)_n$ bridge. Because the competing water trapping of the guanine radical cation ($G^{\cdot+}$) does not depend much upon the number of adjacent A:T base pairs, one expects that in double strands with long $(A{:}T)_n$ sequences between the guanines the

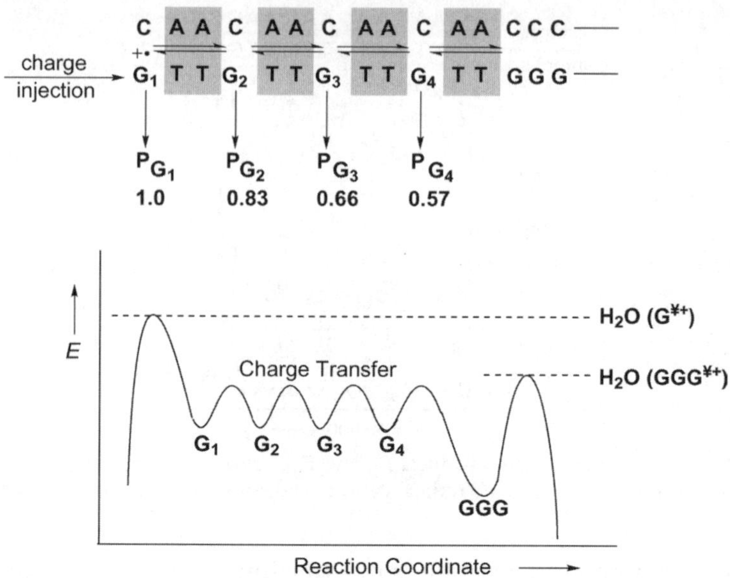

Fig. 11 Yields of the water trapping products P_G at the single guanines, and reaction profile diagram for the electron transfer as well as water trapping of the guanine radical cations

hole transfer will become slower than the trapping reaction by water. This should lead to a collapse of the hole transfer through a DNA, which contains long $(A{:}T)_n$ sequences. But this is not the case, as described in the articles of J.K. Barton and G.B. Schuster in this volume. In order to understand this discrepancy, we performed experiments with an increasing number n of $(A{:}T)_n$ base pairs between the guanine radical cation ($G^{\cdot+}$) and the GGG triplet [17]. For n=1–3 (Fig. 9) the yield ratio P_{GGG}/P_G decreased from 250 to 4.0. But, surprisingly, this ratio changed only very slightly when the number n of the $(A{:}T)_n$ base pairs was further increased (Fig. 12).

These results demonstrate that hole transfer between guanines over short and long $(A{:}T)_n$ sequences follow different distance dependencies and there-

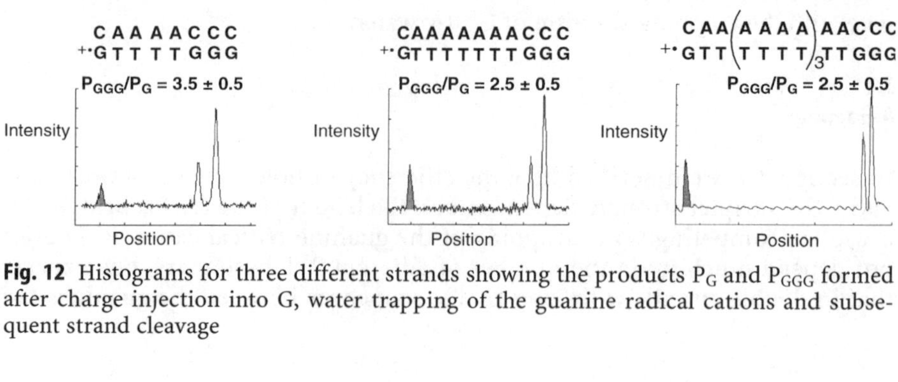

Fig. 12 Histograms for three different strands showing the products P_G and P_{GGG} formed after charge injection into G, water trapping of the guanine radical cations and subsequent strand cleavage

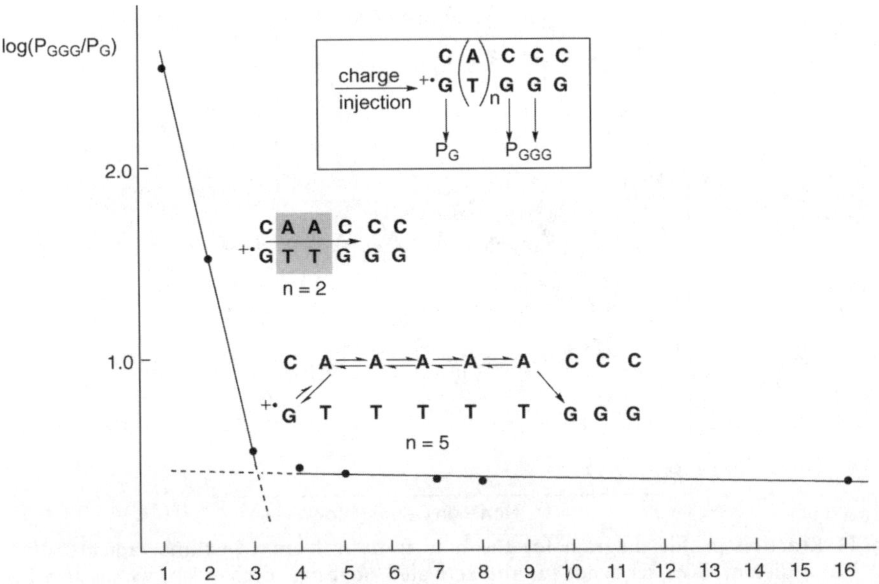

Fig. 13 Influence of the $(A{:}T)_n$ sequence on the hole transfer between a guanine radical cation $G^{\cdot+}$ and a GGG triplet

fore different mechanisms. Figure 13 makes this change of the reaction mechanism visible.

Our explanation is as follows: If the distance between $G^{\cdot+}$ and GGG is short, the electron tunnels in a single step reaction between the guanines, and the $(A{:}T)_n$ sequence acts as a bridge that does not carry the positive charge. But if the distance is long, the tunnelling rate is so slow that an endothermic oxidation of the adjacent adenine becomes faster than the exothermic hole transfer to the GGG triplet. Once the adenine is oxidized, the charge very rapidly travels over the A:T base pairs until it reaches the GGG triplet, the thermodynamic sink for the positive charge (Fig. 14). According to this mechanism the hole transfer over long $(A{:}T)_n$ sequences is possible if the lifetime of the guanine radical cation ($G^{\cdot+}$) is long enough to oxidize the adjacent A:T base pair in a thermally activated step. But under experimental conditions where the lifetime of $G^{\cdot+}$ is too short for this endothermic step, a long $(A{:}T)_n$ sequence can stop the charge transfer through DNA. This is the case for example if the back electron transfer from the injection system reduces the lifetime of $G^{\cdot+}$, as described by K. Nakatani and I. Saito in this volume.

The activated hopping mechanism that also involves adenines as charge carriers is validated by experiments in which we directly injected the positive charge into an adenine [19]. We have shown that photolysis of injection system **18** oxidizes the adenine base, maybe by through-bond charge transfer (**19**→**20**). Therefore, incorporation of **18** into DNA duplexes makes it possible to inject the hole directly into an adenine base in the middle of an

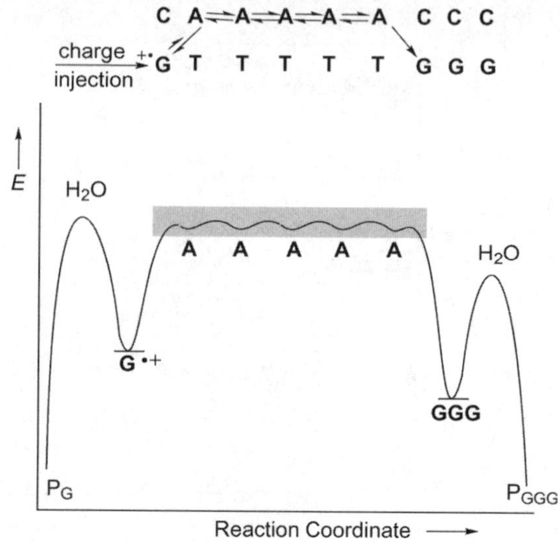

Fig. 14 Reaction profile diagram for the hole transfer from a guanine radical cation ($G^{\cdot+}$) to a distant GGG sequence via the activated hopping mechanism, which also involves adenines (A) as charge carriers

$(A{:}T)_n$ sequence. In double strand **21** (Scheme 5) the charge is trapped by a GGG triplet, and it turned out that the hole transport over 8 A:T base pairs is nearly as efficient as the hole transport over 2 A:T base pairs.

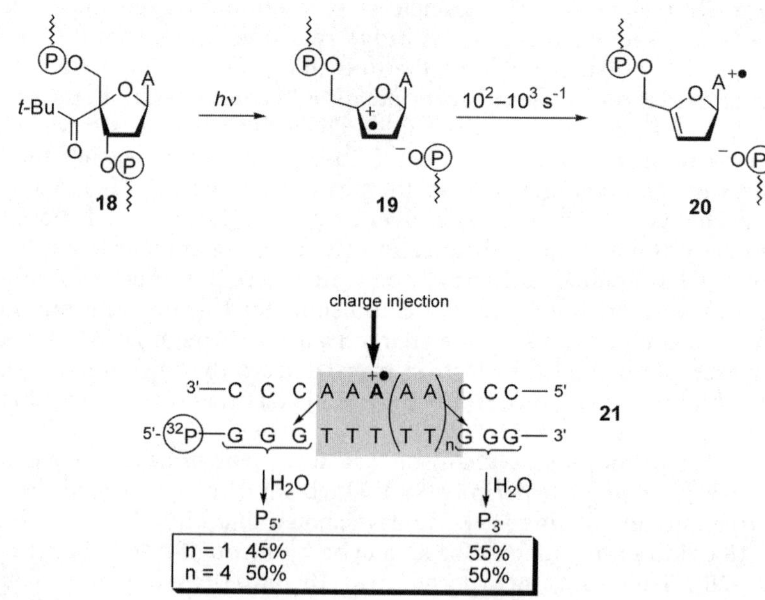

Scheme 5 Charge injection into adenine (A), and efficiency of charge hopping between adenines

This almost distance independent hole transfer over $(A{:}T)_n$ sequences where adenines are charge carriers is very surprising. Maybe the transfer of a positive charge between adenines of an $(A{:}T)_n$ sequence is extremely fast, as recent calculations of M.D. Sevilla predicted [20]. One could also speculate that the positive charge is delocalized over more than one A:T base pair so that polaron hopping, which is discussed in this volume by G.B. Schuster as well as E.N. Conwell, might make the hole transport in oxidized $(A{:}T)_n$ sequences very efficient.

In conclusion, hole transfer between guanines that are separated by long $(A{:}T)_n$ sequences is possible because also the adenines become involved as charge carriers. Such a change of the reaction mechanism can also be explained by calculations [21], and has been described by J. Jortner as well as D. Beratan and M.A. Ratner in their articles of this volume.

4 Mismatches and Proton Transfer

Mismatches can cause a dramatic decrease of the hole transport efficiency, as J.K. Barton and H.H. Holden have shown in their articles of this volume. We have observed these effects in DNA sequences **22–24** (Fig. 15) [18].

In the perfectly paired double strand **22**, the yield of product P_{GGG}, which indicates the amount of charge that has reached the hole trap GGG, is 68%. But if the intermediate G:C base pair is exchanged by a G:T mismatch, the efficiency of the charge transport drops to 23%. With an abasic site (H) opposite to G the hole transport nearly stops at this mismatched site (Fig. 15). We have explained this influence of a mismatch on the efficiency of the charge transport by a proton transfer from the guanine radical cation ($G_2^{\cdot+}$)

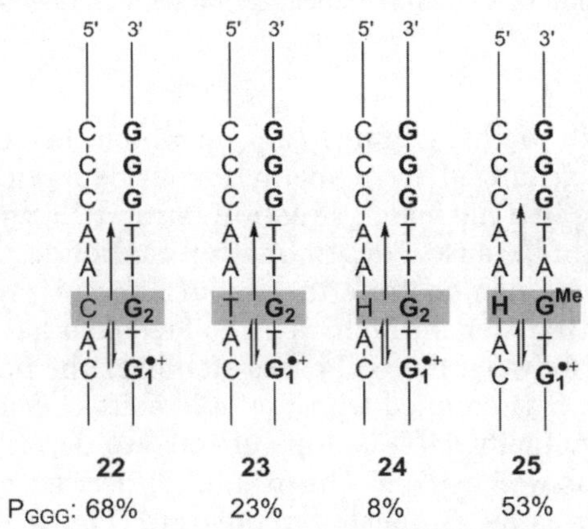

Fig. 15 Influence of mismatches on the efficiency of hole transfer through double strands **23** and **24** where the cytidine (C) is exchanged by thymidine (T) and an abasic site (H), respectively. In **25** guanosine is exchanged by *N*-methylguanosine (see Scheme 6), and C by an abasic site (H)

in the mismatched base pair to the surrounding water. This reaction leads to guanosyl radical **26**, which is a much poorer oxidant than a guanine radical cation ($G^{\cdot+}$), and therefore slows down the charge transport. In perfectly paired G:C base pairs this proton transfer from $G^{\cdot+}$ to water is too slow to compete with the hole transfer because the protons are fixed between the heterocyclic bases (Scheme 6).

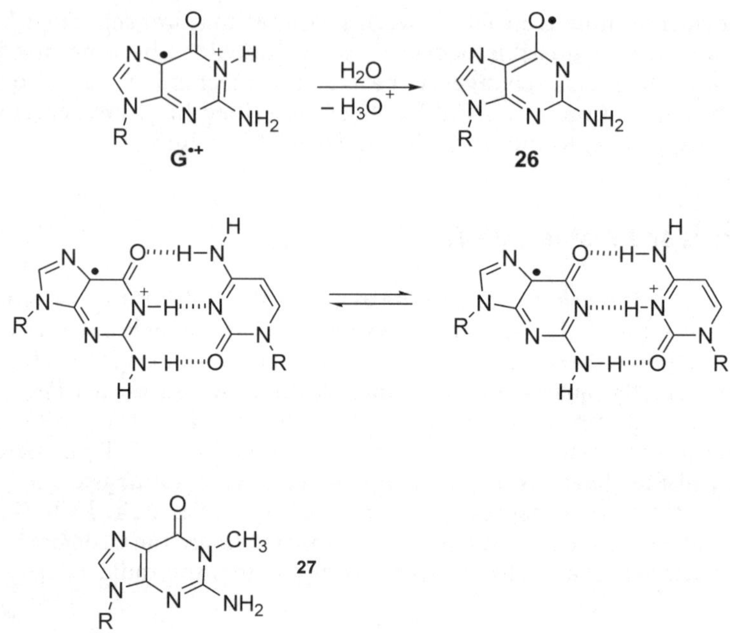

Scheme 6 Deprotonation of a guanine radical cation with and without a complementary base

Experiments with methyl guanine (**27**), in which the acidic proton of the radical cation is exchanged by a methyl group, support this explanation [22]. With this base in a mismatch situation (strand **25**) the hole transfer becomes efficient again because a deprotonation cannot occur (Fig. 15).

These experiments demonstrate the importance of proton transfer processes during hole transfer through DNA. S. Steenken has already remarked that a proton shift between the G:C bases stabilizes the positive charge [23]. If such a proton shift is coupled with the hole shift, a deuterium isotope effect should arise. Actually, H/D isotope effects are described by V. Shafirovich, M.D. Sevilla as well as H.H. Thorp in their articles of this volume. Experiments with our assay [22] also demonstrate (Fig. 16) that hole transfer in protonated DNA (H_2O as solvent) is three times more efficient than in deuterated DNA (D_2O as solvent). If this reflects a primary isotope effect, it shows that the charge transfer is coupled with a proton transfer.

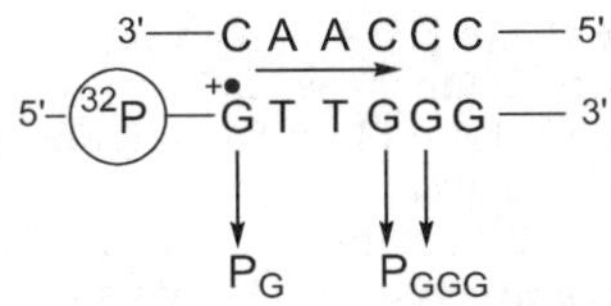

H_2O: $P_{GGG}/P_G = 9.0 \pm 1.0$

D_2O: $P_{GGG}/P_G = 3.0 \pm 1.0$

Fig. 16 Influence of the exchange of protons (H_2O) by deuterons (D_2O) on the efficiency of the hole transfer, measured by the product ratios P_{GGG}/P_G

5 Conclusion

For the discussion of charge transfer in DNA one has to distinguish between the injection of the charge into DNA and the transfer of the charge through DNA. Our experiments, in which a positive charge is site-selectively injected into a guanine, demonstrate that a guanine radical cation ($G^{\cdot+}$) triggers charge transfer through DNA from an adjacent guanine (G). If the $(A:T)_n$ sequence between the guanines is short, the adenines act as a bridge and are not oxidized during the single step hole transfer. Under these conditions an increase of the distance between the guanines decreases the hole transfer rate. Because the reaction does not stop after the first step, long distance hole transport through DNA can occur by a multistep hopping process, using the guanines as charge carriers. The mechanism of the charge transfer changes if the $(A:T)_n$ sequence between the guanines is long. In cases where the lifetime of the guanine radical cation is long enough (for example slow trapping by water), an endothermic oxidation of the adjacent adenine can occur. After this reaction step the positive charge quickly migrates in a hopping process over the adenines until it reaches the next guanine base. This hole transfer process stops when the positive charge is trapped either by nucleophilic attack or by proton transfer to the surrounding water.

A multistep hopping mechanism explains not only the long distance hole transfer through DNA, it can also rationalize the electron transfer through reduced DNA as T. Carell and M.D. Sevilla demonstrate in this volume.

Acknowledgement Our work was supported by the Swiss National Science Foundation and the Volkswagen Foundation.

References

1. Marx A, Erdmann P, Senn M, Körner S, Jungo T, Petretta M, Imwinkelried P, Dussy A, Kulicke KJ, Macko L, Zehnder M, Giese B (1996) Helv Chim Acta 79:1980
2. Giese B, Dussy A, Meggers E, Petretta M, Schwitter U (1997) J Am Chem Soc 119:11130
3. Meggers E, Dussy A, Schäfer T, Giese B (2000) Chem Eur J 6:485
4. Meggers E, Kusch D, Spichty M, Wille U, Giese B (1998) Angew Chem Int Ed 37:460
5. Meggers E (1999) PhD Thesis, University of Basel

6. Giese B, Imwinkelried P, Petretta M (1994) Synlett 1003
7. Giese B, Biland A (2002) Chem Commun 667
8. Steenken S, Jovanovic SV (1997) J Am Chem Soc 119:12950
9. Marcus RA (1956) J Chem Phys 24:966
10. Nakatani R, Dohno C, Saito I (2000) J Am Chem Soc 122:5893
11. Davis WB, Naydenova I, Haselsberger R, Ogrodnik A, Giese B, Michel-Beyerle ME (2000) Angew Chem Int Ed 39:3649
12. Hess S, Götz M, Davis WB, Michel-Beyerle ME (2001) J Am Chem Soc 123:10046
13. Giese B (2000) Acc Chem Res 33:631
14. Meggers E, Michel-Beyerle ME, Giese B (1998) J Am Chem Soc 120:12950
15. Giese B, Spichty M (2000) Chem Phys Chem 1:195
16. Wessely S (2001) PhD Thesis, University of Basel
17. Giese B, Amaudrut J, Köhler AK, Spormann M, Wessely S (2001) Nature 412:318
18. Giese B, Wessely S (2000) Angew Chem Int Ed 39:3490
19. Giese B, Kendrick T (2002) Chem Commun 2016
20. Li X, Cai Z, Sevilla MD (2002) J Phys Chem A 106:9345
21. Segal D, Nitzan A, Davis WB, Wasielewski MR, Ratner MA (??) J Phys Chem B 104:3817
22. Giese B, Wessely S (2001) Chem Commun 2108
23. Steenken S (1997) Biol Chem 378:1293

Top Curr Chem (2004) 236:45–65
DOI 10.1007/b94411

Dynamics and Equilibrium for Single Step Hole Transport Processes in Duplex DNA

Frederick D. Lewis · Michael R. Wasielewski

Department of Chemistry, Northwestern University, Evanston, IL 60208–3113, USA
E-mail: lewis@chem.northwestern.edu
E-mail: wasielew@chem.northwestern.edu

Abstract The dynamics of single step hole transport processes from a guanine cation radical ($G^{+}\cdot$) to GG, GGG and deazaguanine (Z) secondary electron donors separated by one or two A:T base pairs have been investigated by means of nanosecond time-resolved absorption spectroscopy with kinetic modeling. Photoinduced electron transfer from G to a stilbenedicarboxamide (Sa) hairpin linker is used to generate a $G^{+\cdot}/Sa^{-\cdot}$ radical ion pair separated by two or three A:T base pairs. The occurrence of hole transport from $G^{+\cdot}$ to the secondary donors results in an increase in the $Sa^{-\cdot}$ decay time. Kinetic modeling provides both the forward and return rate constants for hole transport, from which the hole transport equilibrium constant and free energy can be obtained. No other experimental method has yielded equilibrium data. Hole transport dynamics depend on the identity of the secondary donor, the number and identity of the bases separating the primary and secondary donor, and the location of the donor within the same strand as the primary donor or in the complementary strand. Secondary GG and GGG donors form very shallow hole traps, whereas Z forms a much deeper hole trap. These results are correlated with other transient absorption studies, selected strand cleavage data, and current theories of charge transport in DNA.

Keywords DNA · Electron transfer · Hole injection · Hole transport · Transient absorption · Kinetic modeling · DNA hairpins

Abbreviations and Symbols

Sa	stilbenedicarboxamide
$Sa^{-\cdot}$	stilbenedicarboxamide anion radical
$G^{+\cdot}$	guanine cation radical
$Z^{+\cdot}$	deazaguanine cation radical
Sd	stilbenediether
$Sd^{+\cdot}$	stilbenediether cation radical

1 Introduction

The possibility that positive charge (holes) might migrate through the π-stacked bases of duplex DNA has long stimulated the imagination of scientists [1, 2]. Evidence for charge migration over short distances was available several decades ago from pulse radiolysis studies of DNA at low temperatures [3]. Recently, evidence for charge migration over several dozen base pairs has been obtained from DNA strand cleavage studies conducted at room temperature [4–6]. Site-selective photooxidation of duplexes possessing multiple GG sequences results in cleavage at sites remote from the initial locus of charge injection [7–9]. Selective cleavage of DNA at guanine-containing sites is attributed to the relative ease of oxidation of G, compared to the other common nucleobases [10]. Even greater selectivity is observed for cleavage at GG and GGG sites, which are believed to function as hole traps [11–15]. Hole migration between G-containing sites has been proposed to occur via a hole-hopping mechanism [16, 17]. The precise mechanism of hole hopping or hole transport and the degree to which holes are delocalized remain topics of active investigation.

Strand cleavage studies have provided relative rate constants for hole transport versus the rate constant for the initial chemical event leading to strand cleavage [18–20]. However, they do not provide absolute rate constants for hole transport processes. Several years ago we introduced a method based on femtosecond time-resolved transient-absorption spectroscopy for investigating the dynamics of charge separation and charge recombination in synthetic DNA hairpins [21, 22]. Recently, we have found that extensions of this method into the nanosecond and microsecond time domains permit investigation of the dynamics of hole transport from a primary hole acceptor to a secondary hole acceptor [23]. Kinetic modeling of this data provides both forward and return hole transport rate constants and hence the equilibrium constants for reversible hole transport. This method provides data about the hole transport process that is unavailable from other experiments.

2 Hole Injection

2.1 Stilbenedicarboxamide-Linked Hairpins

The formation of exceptionally stable synthetic DNA hairpins possessing a stilbene-4,4′-dicarboxamide (**Sa**) linker (Scheme 1) connecting complemen-

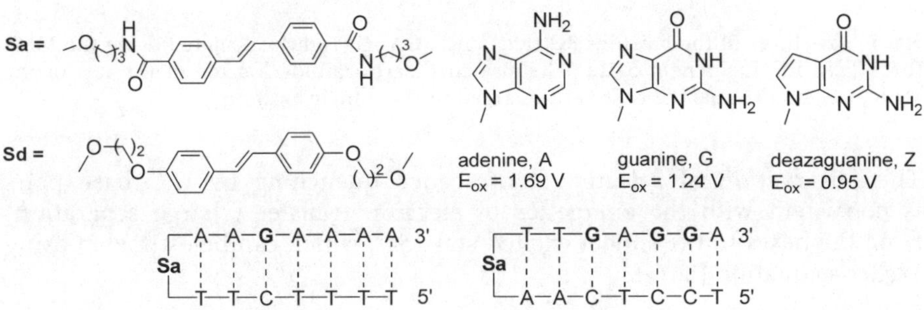

Scheme 1 Structures of the **Sa** and **Sd** hairpin linkers, electron donor nucleobases and their oxidation potentials, and the hairpins **3G** and **3GAGG** shown in Fig. 1

tary oligonucleotide "arms" was reported by Letsinger and Wu in 1995 [24]. The assignment of a B-DNA structure in which the stilbene linker is π-stacked with the adjacent base pair was proposed on the basis of spectroscopic data and later supported by x-ray crystallography for an analogous hairpin possessing a stilbenediether linker (**Sd**) [25]. MM^+ minimized structures for two **Sa**-linked hairpins possessing either a single G:C base pair (**3G**) or three G:C base pairs (**3GAGG**) are shown in Fig. 1. Condensed formulae for these hairpins are shown in Scheme 1. The preparation of these hairpins by standard phosphoramidite chemistry using a mono-protected, mono-activated chromophore has facilitated the synthesis of dozens of synthetic hairpins possessing **Sa** and other linkers and oligonucleotide arms of varying length and base sequence.

Letsinger and Wu [24] also made the seminal observation that **Sa**-linked hairpins possessing polyT and polyA arms are strongly fluorescent, whereas the **Sa** fluorescence is strongly quenched by a neighboring G:C base pair.

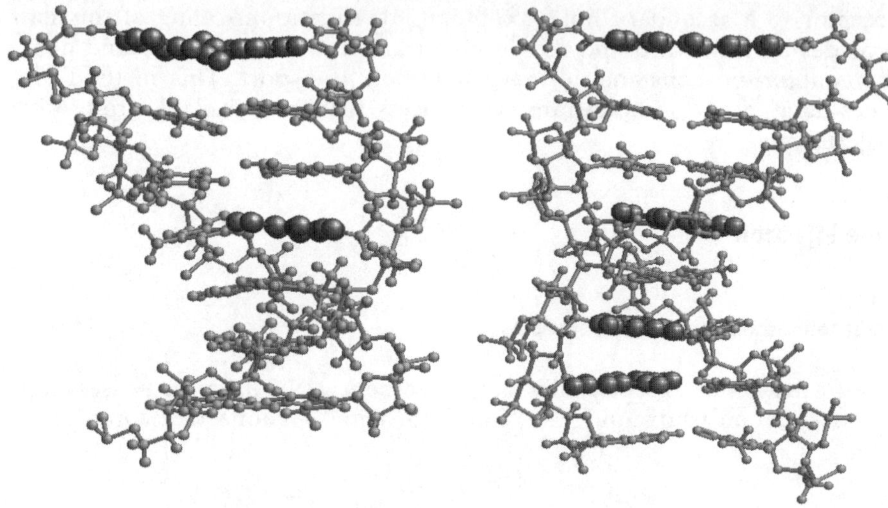

Fig. 1 Structures of the hairpins **3G** (left) and **3GAGG** (right) calculated using the MM^+ force field in Hyperchem 5.01a. The stilbenedicarboxamide located at the top of the structure and the guanine bases are enlarged to show their location

The observation of selective fluorescence quenching by G:C base pairs is consistent with the energetics of electron transfer (charge separation) from the bases to the singlet excited state **Sa***, which can be estimated using Weller's equation [26]:

$$\Delta G_{CS} = -(E_s + E_{rdn}) + E_{ox} \tag{1}$$

where E_s is the **Sa** singlet energy (3.35 eV), E_{red} is its reduction potential (−2.05 V vs SCE), and E_{ox} is the nucleobase oxidation potential (Scheme 1). Lacking measured potentials for the bases in duplex DNA, we have used the single-nucleotide values of Seidel et al. [10], which were determined in polar, aprotic solvents. As shown in Fig. 2, the values of ΔG_{cs} for photooxidation of G and A are −0.3 and +0.2 eV, respectively. Oxidation of T or C is even more endergonic, in accord with selective quenching of **Sa*** by G. Also evident from Fig. 2a is the highly exergonic nature of the charge recombination process.

2.2 Dynamics of Charge Separation and Charge Recombination

The dynamics of photoinduced charge separation, k_{cs}, and charge recombination, k_{cr} (Fig. 2a), have been studied in several families of hairpins containing an **Sa** linker and a single G:C base pair by means of femtosecond time-resolved transient absorption spectroscopy [27, 28]. Both the singlet state and anion radical of **Sa** have strong transient absorption centered at 575 nm. The difference in the independently determined band shapes for **Sa***

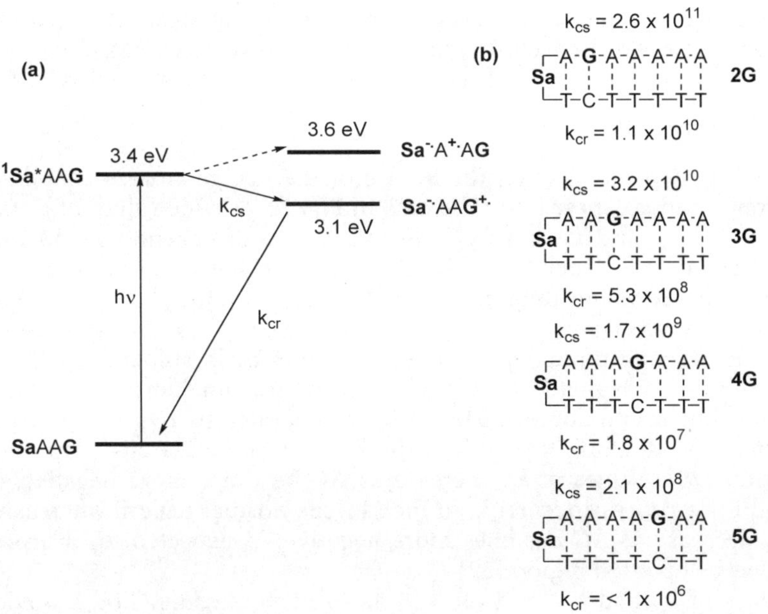

Fig. 2 a Energetics of photooxidation of G and A by singlet **Sa**. **b** Dynamics of charge separation (k_{cs}) and charge recombination (k_{cr}) for **Sa**-linked hairpins possessing a single guanine

and **Sa**$^{-\cdot}$ makes it possible to determine the decay times for these species from transient spectra such as those shown in Fig. 3 for the hairpin **3G**.

Values of k_{cs} and k_{cr} for a family of hairpins containing a single G:C base pair separated from **Sa** by one-to-four A:T base pairs are summarized in

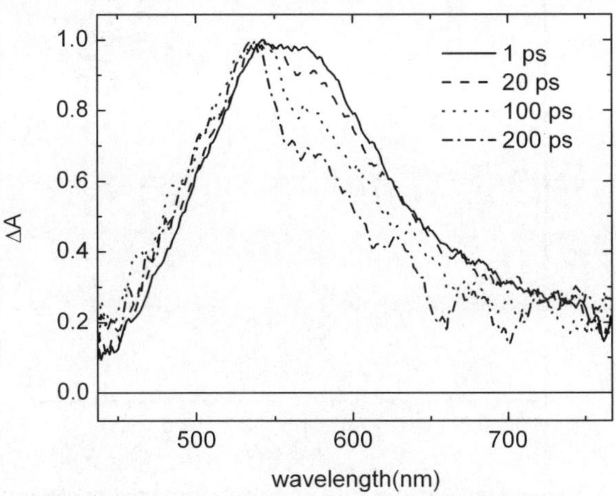

Fig. 3 Transient absorption spectra of hairpin **3G** obtained at increasing delay times following 340 nm excitation with a laser system having a 150 fs instrument response function

Fig. 2b. Both k_{cs} and k_{cr} are seen to decrease as the distance R between **Sa** and G:C increases. In accord with a superexchange mechanism for photoinduced electron transfer, the distance dependence can be described by Eq. (2):

$$k_{et}=k_o \exp(-\beta R) \qquad (2)$$

where β is dependent upon the nature of the bridge and its coupling with the donor and acceptor [29]. The data in Fig. 2b provide values of β=0.7 and 0.9 Å^{-1} for k_{cs} and k_{cr}, respectively. We have subsequently found that the magnitude of β is dependent upon the donor-bridge-acceptor energetics, recently obtained values ranging from 0.4 to 1.1 Å^{-1} [30].

Time-resolved transient absorption spectroscopy has also been utilized to study the driving force dependence of k_{cs} and k_{cr} in stilbene- and phenanthrene-linked hairpins [31]. A plot of k_{et} vs ΔG_{et} for hairpins in which the acceptor linker and donor nucleobase are separated by two A:T base pairs is shown in Fig. 4, along with a fit of the k_{cs} data to the Marcus-Levich-Jortner equation [29]. Values of k_{cs} increase as ΔG becomes more negative, as expected for a process occurring in the Marcus normal region, whereas values of k_{cr} decrease as ΔG becomes more negative, as expected for a process in the Marcus inverted region [32].

Values of k_{cs} and k_{cr} for **Sa**-linked hairpins possessing G or Z single base donors and GG or GGG sequences, which are expected to be more readily oxidized than G, are summarized in Fig. 5. Values of k_{cs} for Z, GG, and GGG donors are slightly faster than that for G. Similar results for GG and GGG have been reported by Davis et al. [33]. The small variation in k_{cs} may reflect the moderately exergonic nature of these processes, which may place them near the maximum in the Marcus curve (Fig. 4), resulting in a kinetic leveling effect. In fact, Sistare et al. [34] report a value of k_{GG}/k_G=12, based on

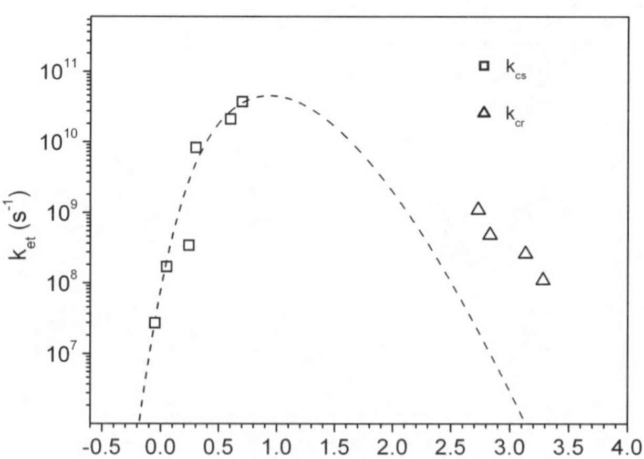

Fig. 4 Free energy dependence of the rate constants for charge separation and charge recombination for hairpins in which two A:T base pairs separate the linker acceptor from the nucleobase donor. The dashed line is a fit of the charge separation data to the Marcus-Levitch-Jortner equation

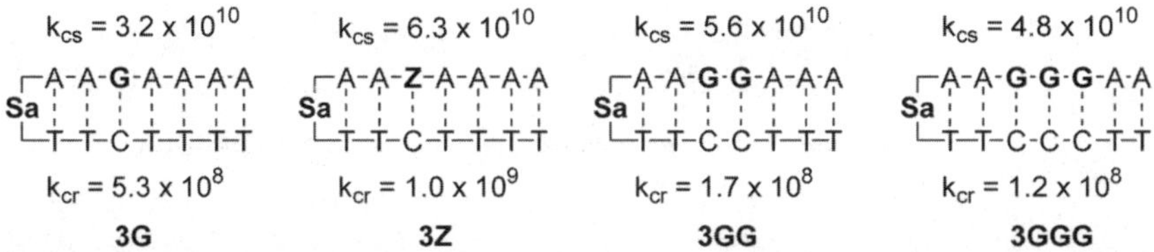

Fig. 5 Dynamics of charge separation and charge recombination for hairpins possessing G, Z, GG, and GGG donors

their electrochemical measurements. The value of k_{cr} for Z is larger than that for G, in accord with less exergonic electron transfer in the inverted region. However, the values for GG and GGG are smaller than that for G. This may reflect either hole delocalization over the GG or GGG sequence or equilibration of a localized hole. The dynamics of charge separation between strong acceptor linkers and neighboring A:T vs G:C base pairs indicates that the difference between A and G oxidation potentials is large [31, 35], consistent with the potentials reported by Seidel et al. [10].

3
Hole Transport Across a Single A:T Base Pair

3.1
General Considerations and Kinetic Modeling

The design concept for the study of hole transport from a primary guanine donor to a secondary deazaguanine donor in the **Sa**-linked hairpin **3GAGG** is shown in Scheme 2 and the full kinetic scheme in Fig. 6. The occurrence

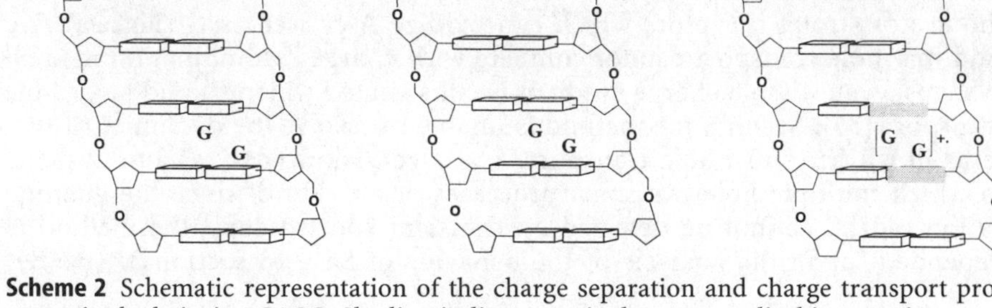

Scheme 2 Schematic representation of the charge separation and charge transport processes in the hairpin **3GAGG**. Shading indicates excited state or radical ion species

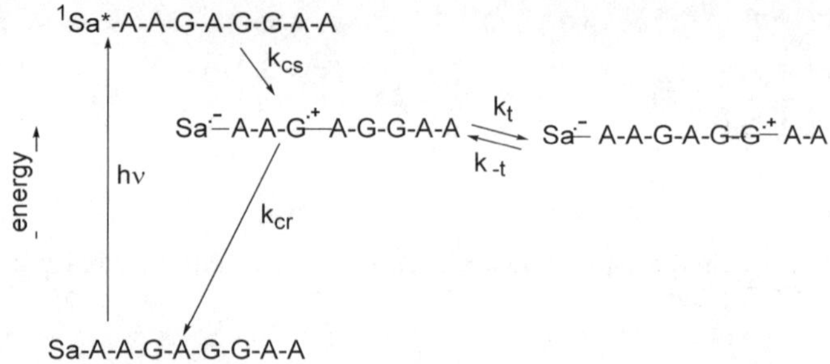

Fig. 6 Kinetic scheme for reversible charge transport in a GAGG sequence

of hole transport requires that the rate constant for charge transport, k_t, competes with charge recombination, k_{cr}, in the initially-formed $\mathbf{Sa}^{-\cdot}$-$\mathbf{G}^{+\cdot}$ radical ion pair. The occurrence of charge transport should result in a long-lived $\mathbf{Sa}^{+\cdot}$ decay component. If $k_t \sim k_{cr}$, then two decay components should be observed, their relative amplitudes reflecting the relative rate constants for the two competing decay processes of the $\mathbf{Sa}^{-\cdot}$-$\mathbf{G}^{+\cdot}$ radical ion pair. The shorter-lived component should have a decay time shorter than that of a hairpin lacking a secondary donor. The longer-lived component should have a decay time longer than that of the hairpin lacking a secondary donor and might not decay on the μs time scale of our experiments if either return hole transport, k_{-t}, is very slow or $\mathbf{GG}^{+\cdot}$ undergoes a relatively fast chemical reaction.

The ps and ns transient decays of hairpin **3GAGG** are shown in Fig. 7. The ps decay time is similar to that for hairpin **3G**, in accord with similar hole injection rates for the two hairpins. The nanosecond decay of **3GAGG** is biexponential, the short-lived component having a shorter decay time than that for hairpin **3G** and the second component having a significantly longer decay time. Kinetic modeling of the transient decays provides values for both k_t and k_{-t}, and hence the equilibrium constant for hole transport, K_{ht}, from which the free energy change, ΔG_{ht}, can be calculated.

Whereas other experimental methods have been used to obtain values of k_t, no other method provides values of k_{-t} or equilibrium data. There are, however, several important limitations of our method. First, the method is restricted to relatively fast hole transport processes that can compete with charge recombination of the $\mathbf{Sa}^{-\cdot}$-$\mathbf{G}^{+\cdot}$ radical ion pair (Fig. 6). This precludes the use of strong acceptors which can oxidize A as well as G (Fig. 2a). We find that hole transport cannot compete with charge recombination in such systems, even when a charge gradient is constructed which should favor hole transport [35]. Second, the method is unable to resolve the dynamics of systems in which return hole transport, k_{-t}, is very slow ($<10^4$ s^{-1}) or systems in which multiple hole transport processes occur. Third, since the guanine cation radical cannot be detected by transient spectroscopy, the method is dependent upon the analysis of the behavior of $\mathbf{Sa}^{-\cdot}$. In section 3.4 we de-

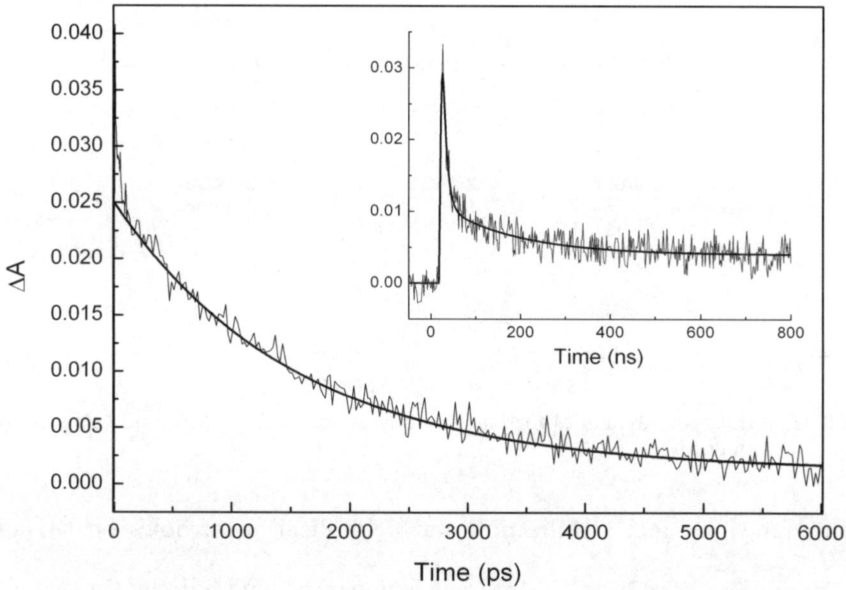

Fig. 7 Transient decay for the hairpin **3GAGG** on the ps and ns (inset) time scales. Fits to the data were obtained using the Levenberg-Marquardt algorithm

scribe a promising new system in which the donor cation radical as well as the acceptor anion radical are observed.

3.2 Hole Transport in a GAGG, GAGGG, and GAZ Systems

The transient absorption spectrum of **3GAZ** displays a long lived **Sa**$^{-\cdot}$ component with a large amplitude [36]. Therefore charge transport competes effectively with charge recombination in this system. Kinetic analysis provides values of k_t=6.9×10^8 s^{-1} and k_{-t}=4×10^5 s^{-1} (Fig. 8), from which values of K_{ht}=1,700 and ΔG_{ht}=−0.19 eV can be calculated (Table 1). The calculated value for ΔG_{ht} is somewhat smaller than the measured difference in irreversible oxidation potentials of dZ vs dG (ca. 0.3 eV) [37] or calculated ionization potentials of Z vs G (0.39 eV) [12]. Computational studies indicate that neighboring bases in B-DNA can influence the effective base ionization potentials. The smaller value of G_{ht}=−0.19 eV may, in fact, reflect a "leveling effect" of the neighboring bases on the relative energies of Z$^{+\cdot}$ vs G$^{+\cdot}$ [38].

GG and GGG sequences have been widely used in strand cleavage studies of hole migration in DNA [4–6]. According to conventional wisdom, GG and GGG sequences serve as "hole traps". The calculated gas phase ionization potentials reported by Sugiyama and Saito [14] provide relative energies for G vs GG (0.47 eV) and G vs GGG (0.68 eV) that continue to be cited as evi-

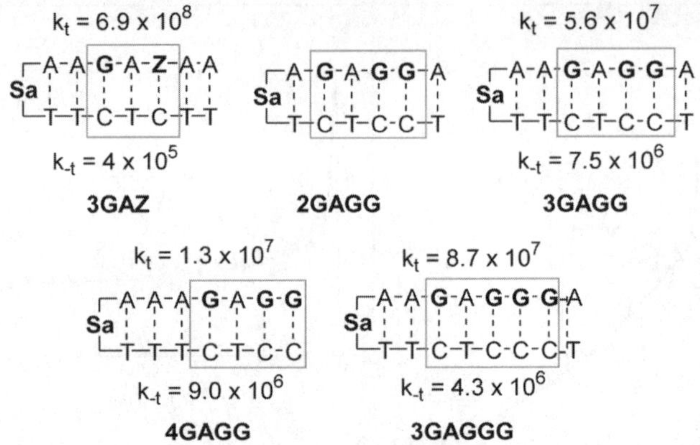

Fig. 8 Hole transport dynamics for hairpins possessing GAZ, GAGG, and GAGGG hole transport sequences

dence for significant stabilization and delocalization of holes on GG and GGG.

The ns decay for **2GAGG** is a single exponential with a decay time similar to that for **2G** (Fig. 2b) [23, 39]. This means that there is no evidence for the occurrence of hole transport in **2GAGG.** In contrast, dual exponential ns decay is observed for **3GAGG.** Kinetic modeling provides values of k_t=5.6×10^7 s^{-1} and k_{-t}=7.5×10^6 s^{-1} (Fig. 8), from which values of K_{ht}=7.5 and ΔG_{ht}=−0.052 eV can be calculated (Table 1). The value of k_t is much smaller than the value of k_{cs} for **2GAGG,** accounting for the absence of measurable charge transport in that sequence. Similar values of ΔG_{ht} were obtained for hairpins with **Sa-AAAGTGGA** and **Sa-TTTGAGGA** donor sequences. A slightly larger value of ΔG_{ht}=0.077 eV was obtained for **3GAGGG,** indicative of the formation of a slightly more stable hole trap (Table 1).

Table 1 Equilibrium constants and free energy changes for hole transport

Hairpin[a]	10^{-7} k_t (s^{-1} [b])	K_{ht}=k_t/k_{-t} [c]	−ΔG_{ht} (eV[d])
3GAZ	69	1,700	0.19
4GAGG*	1.3	1.4	0.008
3GAGG	5.6	7.5	0.052
3GAGGG	8.7	20	0.077
4GTGG	0.04	6.7	0.05
4GACC*	0.21	5.7	0.04

[a] See Fig. 8 and Fig. 9 for conjugate structure and donor strand sequence. *indicates terminal C:G base pair

[b] Rate constant for hole transport from Fig. 8 and Fig. 9

[c] Equilibrium constant for hole transport, calculated using experimental values of k_t and k_{-t} from Fig. 8 and Fig. 9

[d] Free energy change for hole transport. ΔG_{ht}=−RT [ln(K_{ht})]

Smaller values of ΔG_{ht}~0.01 eV were obtained for **4GAGG** and for a hairpin with a **Sa-TTTGAGG** donor sequence. Evidently, a terminal GG base pair has approximately the same energy as an internal G in a AGA sequence. Thus we suggest that the value of k_t for **4GAGG** provides an approximate value of 1×10^7 for isoenergetic hole transport in a GAG sequence.

The values of k_t for hole transport from the secondary donor to the primary donor in Fig. 8 are substantially smaller than the values of k_{cs} for the hairpins in Fig. 5. The slower rate constants for hole transport vs charge separation, which occurs at a longer donor-acceptor distance, may reflect fundamental differences between these two processes [40]. The charge separation process occurs via a superexchange mechanism with a relatively modest solvent reorganization energy [31]. To our knowledge, the temperature dependence of hole transport dynamics have not been determined. However, there is growing evidence that hole transport may be gated by large amplitude motions of the nucleobases, solvent, and ions [40].

The modest variation in k_t values for the hairpins **4GAGG, 3GAGG,** and **3GAGGG** (k_t(rel)=1:4:7) is consistent with the relatively small values of ΔG_{ht} for hole transport from G to GG or GGG (Table 1). In contrast, the value of k_t for hairpin **3GAZ** is much larger than that for any of the hairpins with G secondary donors, in accord with the larger value of ΔG_{ht} for **3GAZ**. The variation in k_t with the secondary hole acceptor is larger than the variation in k_{cs} for the same set of acceptors (Fig. 5), which we attribute to a kinetic leveling effect for electron transfer processes which occur near the top of the Marcus curve.

3.3
Pathway Dependence in G-B-GG Systems

Values of k_t and k_{-t} have been determined for several **GAGG** systems in which the primary G donor is separated from the **Sa** acceptor by a T_n rather than an A_n bridge [41]. Whereas the values of k_{cs} are somewhat slower for charge separation via T_n vs A_n bridges, there is little difference in the values of k_t or k_{-t} for these bridges. In contrast, systems with **GTGG** hole transport sequences have much smaller values of k_t than do systems with **GAGG** sequences. Kinetic data for the **4GTGG** system is shown in Fig. 9 and equilibrium data reported in Table 1. Even though the rate constants for this and related systems are 50–100 fold slower than for **GAGG** systems (Fig. 8), the equilibrium constant remains essentially the same. Attempts to detect hole transport in several other **GTGG** systems were unsuccessful, presumably due to the failure of slow charge transport to compete with charge recombination.

The dynamics of interstrand hole transport have also been investigated for several hairpins possessing **GACC** hole transport sequences in which a GG secondary donor is located in the complementary strand [41]. Kinetic data for **4GACC** are reported in Fig. 9 and the resulting equilibrium data in Table 1. Comparison of the values of k_t for **4GACC** and **4GAGG** shows that there is a "kinetic penalty" of 1/6 for inter- vs intrastrand hole transport. A

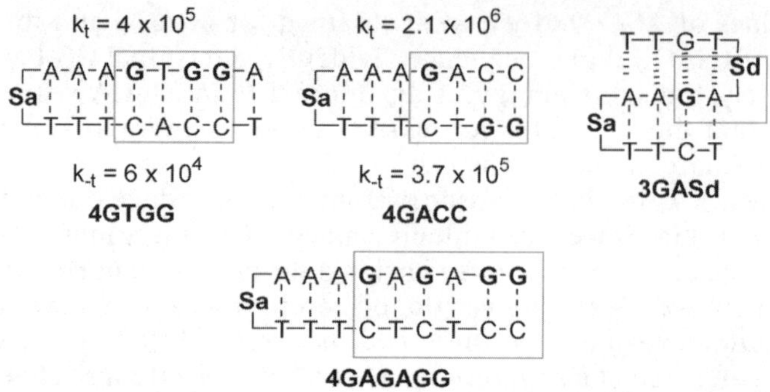

Fig. 9 Hole transport dynamics and structures for hairpins possessing GTGG and cross-strand GACC hole transport sequences, a triplex hole transport system, and a two-step hole transport system

similar result was obtained for analogous sequences in which **Sa** and the donors are separated by a T_3 bridge.

We have also investigated the transient spectroscopy of hairpins such as **4GAGAGG** (Fig. 9), which possess two hopping steps [40]. While a long lived-component of **Sa**$^{-\cdot}$ transient decay can be observed, it is not possible to obtain the rate constants from a fit of the transient decay curves.

3.4 Hole Transport in a Triplex GASd Sequence

We have recently investigated electron transfer processes in DNA triplex systems in which the Watson-Crick base pairs are linked by **Sa** and the Hoogsteen base pairs by a stilbenediether (**Sd**) linker [42]. The **Sd** linker has a lower oxidation potential than does G and so can serve either as a primary donor in studies of charge separation and charge recombination, or as a secondary donor in studies of hole transport. The triplex **3GASd** (Fig. 9) possesses a guanine primary donor and an **Sd** secondary donor. The transient decay of **Sd**$^{-\cdot}$ in **3GASd** displays dual exponential decay, indicative of the occurrence of reversible hole transport from G to **Sd**. The long-lived component has a decay time similar to that of **3GAZ**, indicative of a large equilibrium constant for hole transport from G to **Sd**. Since the transient absorption spectra of **Sa**$^{-\cdot}$ and **Sd**$^{+\cdot}$ can be resolved, it may be possible to distinguish the rise time of these two species and directly determine the hole transport rates.

4 Distance Dependence of Hole Transport Dynamics

Several hairpins containing GAAGG hole transport sequences were designed to investigate the dynamics of hole transport across an AA bridge [41]. Only in the case of **4′GAAGG** (the primes refer to hairpins in which **Sa** and the

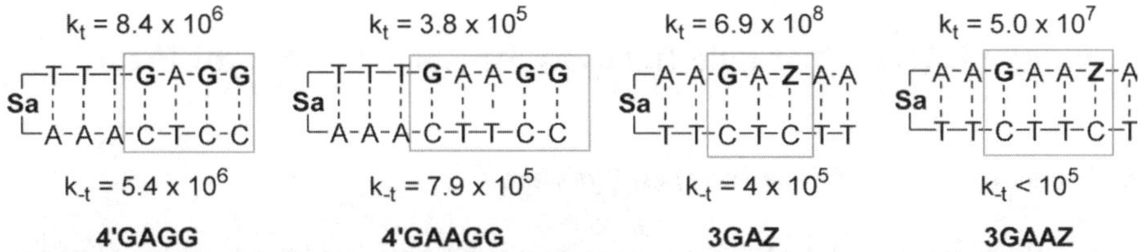

Fig. 10 Hole transport dynamics for hairpins with GAAGG and GAAZ hole transport sequences

donor sequence are separated by a T_n bridge), which has a very slow rate constant for charge recombination, was it possible to detect a long-lived **Sa**$^{-\cdot}$ decay component. Kinetic modeling provides values of k_t and k_{-t} which are reported in Fig. 10, along with data for **4′GAGG**. The value of k_t for **4′GAAGG** is 22-fold slower than that for **4′GAGG**. The difference in values for k_{-t} is smaller, but this data is less reliable. In both cases, values of K_{ht}~1 are obtained, as previously noted for **4GAGG** (Fig. 8), which also has a terminal G:C base pair.

Hole transport in conjugate **3GAAZ** might be expected to be more rapid and exergonic than for a GAAGG sequence. A long-lived decay component was observed for conjugate **4GAAZ**; however, it did not decay appreciably on the μs time scale of our measurements [36]. Kinetic modeling provided an estimated value of k_t=5.0×10^7 s^{-1}, appreciably faster than the value for **4′GAAGG**, but 14-fold slower than the value for **3GAZ** (Fig. 10). Our limited data for the distance dependence of k_t is shown in Fig. 11.

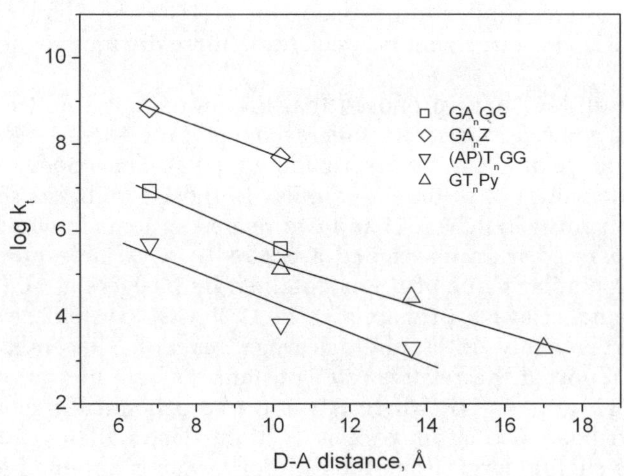

Fig. 11 Distance dependent kinetics of hole transport in GA_nGG, GA_nZ, $(AP)T_nGG$ [43] and GT_nPy [44] systems

5
Comparisons with Other Hole Transport Experiments and with Theory

5.1
Experimental Studies of Hole Transport Dynamics

Two other studies of hole transport dynamics have been reported. Shafirovich et al. [43] have investigated the dynamics of hole transport in $[AP^{+\cdot}]T_nGG$ systems (where $AP^{+\cdot}$ is the aminopurine cation radical generated by two-photon ionization), detecting guanine radical formation by transient absorption spectroscopy on a microsecond time scale. Their data for n=1, 2, and 3 are shown in Fig. 11. Our value of k_t for **4GTGG** is in excellent agreement with their value for n=1. They also investigated $[AP^{+\cdot}]A_nGG$ systems. Whereas they were unable to resolve the hole transport dynamics for n=1–3, their value of k_t for an A_4 bridge is ca. 10^4-fold faster than for a T_4 bridge. This difference is larger than we observe for a A_1 vs T_1 bridge. However, both studies indicate that A_n is a much more effective bridge for charge transport than T_n. These results stand in marked contrast with our earlier studies of charge separation, in which similar values of k_{cs} were observed for **Sa**-A_nG sequences (Fig. 2b) and for **Sa**-T_nG sequences [27].

The distance dependence of hole transport in $G^{+\cdot}T_nPy$ systems, where Py is a pyrene secondary donor separated from a G-rich region of DNA by a T_n bridge, has been investigated by Kawai et al. [44]. Pulse radiolysis was used to generate guanine cation radicals, and the formation of $Py^{+\cdot}$ was studied by transient absorption spectroscopy on the microsecond time scale. Data for n=2, 3, and 4 is shown in Fig. 4. The values of k_t for n=2 and 3 are somewhat faster than the values of Shafirovich et al. [43] although the extrapolated value for n=1 is similar to our value for **4GTGG** and Shafirovich's value for $[AP^{+\cdot}]TGG$. The agreement between these three distinctive methods is remarkable.

Jortner et al. [16] have proposed that the individual hole transport processes can be viewed as discrete superexchange processes. As such, the distance dependence of k_t can be described by Eq. (2). The slopes of the data in Fig. 4 provide values of $\beta \sim 0.6$ Å^{-1}, similar to those for charge separation in **Sa**-linked hairpins (Fig. 2b) [27] and two related systems in which a guanine donor is separated from an excited acceptor by a variable number of A:T base pairs. A similar value of β was obtained by Meggers et al. [17] in their study of strand cleavage probabilities in $G^{+\cdot}T_nGGG$ systems, where $G^{+\cdot}$ is generated irreversibly in a photocleavage reaction. Recently Giese and Biland [20] reported the relative rate constant for hole migration in the sequences $R^{+\cdot}TZ$ and $R^{+\cdot}TG$, where $R^{+\cdot}$ is a photochemically generated oxidized deoxyribose. Oxidation of Z is 17 times faster than oxidation of G. This value is slightly larger than the ratio of k_t values obtained for **3GAZ** vs **3GAGG** (Table 1). The agreement between our kinetic data and Giese's strand-cleavage data is reassuring.

To our knowledge, there are no other direct measurements of the dynamics of interstrand hole transport. However, the strand cleavage studies of Meggers et al. [17] have clearly demonstrated that long-distance hole transport can occur via a G-hopping sequence in which guanines are located in both strands. Other workers have assumed that hole transport occurs exclusively via intrastrand pathways [45].

5.2 Theoretical Studies of Hole Transport Dynamics

Theoretical studies of hole transport dynamics in DNA far outnumber the experimental studies summarized above. Prior to our initial report of the hole transport rates for the GAGG sequence, Bixon et al. [46] estimated a value of k_t~10^9 s^{-1} for hole hopping from G to G via two intervening AT base pairs. Both the motions of polarons through DNA and incoherent hole transport in DNA have been estimated to occur on the ps time scale [47, 48]. These estimated rates are comparable to our measured value for superexchange charge separation in the **Sa**AAG sequence (k_{cs}=3×10^{10} s^{-1}), but are much faster than the value of k_t=4×10^5 s^{-1} for hole transport in a GAAGG sequence (Fig. 10) or than our estimate for isoenergetic GAG hole transport, k_t~10^7 s^{-1}. Evidence from both theory and experiment suggests that the relatively low values for k_t may reflect the presence of solvent-induced barriers for hole transport or gating by the motions of the duplex or cations associated with the duplex [49].

There has been considerable recent interest among theoreticians in the distance dependence of the dynamics of hole transport. For $G(A)_nG$ sequences, Jortner and co-workers [50, 51] have proposed that hole transport from G to G occurs via a single step superexchange process when $n\leq3$. Similar results have been obtained by Berlin et al. [52, 53]. For longer A_n sequences these workers propose a change in mechanism to thermally induced hopping in which the A_n bridge is oxidized. Since our results pertain only to short distance hole transport, the latter mechanism does not seem to be relevant to our studies. Troisi and Orlandi [54] have calculated values of β~0.7 Å for hole transport in $G(A)_nG$ systems using the electronic couplings between adjacent bases obtained from an ab initio procedure. A significantly higher value of β~1.5 Å was obtained by Olofsson and Larsson [55] using electronic couplings obtained from a superexchange model. The lower value of β is in good agreement with the data in Fig. 11.

Theoreticians have also considered the difference in the distance dependence of hole transport in $G(T)_nG$ sequences. Voityuk et al. [56] predict more rapid hole transport for T_n vs A_n bridges and a ca. 10-fold decrease in k_t for each additional bridging base. Olofsson and Larsson [55] also predict more rapid hole transport via T_n vs A_n bridges. However, Troisi and Orlandi [54] and Rak et al. [57] have recently reached the opposite conclusion, going so far as to predict that hole transport in $G(T)_nG$ sequences would occur via the polyA sequence in the complementary strand. If hole transport does in fact occur via the complementary A_n bridge in T_n-bridged systems, then the value of β should be similar to that for systems with A_n bridges. Comparison of our data for GA_nGG or GA_nZ systems with the data of Shafirovich et al. [43] and Kawai et al. [44] for T_n bridges indicates that this appears to be the case.

The dynamics of inter- vs intrastrand hole transport has also been the subject of several theoretical investigations. Bixon and Jortner [38] initially estimated a penalty factor of ca. 1/30 for interstrand vs intrastrand G to G hole transport via a single intervening A:T base pair, based on the matrix elements computed by Voityuk et al. [56]. A more recent analysis by Jortner et al. [50] of strand cleavage results reported by Barton et al. [45] led to the proposal that the penalty factor depends on strand polarity, with a factor of 1/3 found for a 5′-GAC(G) sequence and 1/40 for a 3′-GAC(G) sequence (interstrand hole acceptor in parentheses). The origin of this penalty is the reduced electronic coupling between bases in complementary strands.

Our experimental data (Fig. 9) provide penalties of 1/6 for inter- vs intrastrand 5′-GACC(GG) hole transport. In view of the difference between our "mini-hairpins", which contain a single hole transport step and Barton's duplex, which contains multiple hopping steps, the agreement between our penalty and those reported by Jortner et al. [50] is excellent. The small penalty for inter- vs intrastrand hole transport lends credence to the proposals of Rak et al. [57] and Troisi and Orlandi [54] that hole transport in $G(T)_n$ sequences may occur via the complementary A_n sequence. We, in fact, observe a much larger kinetic penalty for GTGG vs GAGG hole transport than for interstrand GACC(GG) hole transport. If two strand crossings are involved in this process, the average kinetic penalty for each crossing would be ca. 1/10, in good agreement with our value of 1/6 for a single strand crossing.

5.3
Experimental Studies of Hole Transport Equilibria

Our kinetic modeling method provides the only experimental values for k_{-t} as well as k_t and hence the only available values of K_{ht} and ΔG_{ht} (Table 1). The small stabilization energies for holes located on GG or GGG vs G (Table 1) are consistent with experimental strand cleavage data for short duplexes containing a limited number of G, GG, and GGG sites, which display a relatively low level of strand cleavage selectivity [9, 11, 13, 15]. Data from several such studies provides a "consensus" ratio of 1:3:5 for cleavage at G vs GG vs GGG. Therefore cleavage is less selective than would be expected on the basis of the equilibrium ratio of 1:7.5:20 provided by our equilibrium data (Table 1). The ratio of relative cleavage yield to relative population provides a measure of the relative reactivity on a per-site basis of 1:0.4:0.2. Giese and Spichty [58] arrive at a similar ratio, 1:0.7:0.3, based on their analysis of strand cleavage ratios for several duplexes. The agreement between these results, which were obtained by totally different methods, is reassuring. Bixon and Jortner [59] obtain a somewhat larger ratio for G vs GGG,

Table 2 Comparison of G, GG, and GGG sequences: Relative populations, cleavage yields, and reactivities on a per-sequence and per-G basis

Property of sequence	G	GG	GGG
Population[a]	1	7.5	20
Cleavage yield[b]	1	3	5
Cleavage rate[c]	1	0.4	0.2
Cleavage rate per G[d]	1	0.2	0.1

[a] Relative equilibrium population based on equilibrium data in Table 1
[b] Consensus value for the relative cleavage yields in sequences containing G, GG, and GGG sites (see text)
[c] (cleavage yield)/population
[d] (cleavage yield)/(population × number of G's per site)

1:0.6 based on their analysis of Giese's data. The relative reactivity on a per-G basis is, of course, even lower.

Our data provide a ratio of 1:0.2:0.1 for reactivity at G, GG, and GGG sites on a per-G basis. These relationships are summarized in Table 2. The small difference in G vs GG vs GGG hole stability and lower reactivity of the more stable sites is, of course, precisely the combination of stability and reactivity that is required for the observation of hole transport over long distances containing multiple G and GG sites. Hole transport between two GGG sites separated by a TTGTT sequence appears to be somewhat slower than cleavage at the initially oxidized GGG site [12]. To our knowledge, the efficiency of strand cleavage in duplexes possessing more than two GGG sites has not been studied.

Kinetic modeling provides a significantly larger value of $\Delta G_{ht}=-0.19$ eV for hole transport in conjugate **3GAZ** (Table 1). Therefore Z functions as a much deeper hole trap than does GG or GGG. The larger value of ΔG_{ht} for Z vs GG is also consistent with the faster rate constants for both superexchange charge separation [36, 60] and relative rates determined from strand cleavage studies by Giese and Biland [20] for hole transport from an oxidized sugar to Z vs GG [61]. Nakatani et al. [12] have reported that hole injection into the 5′-end of a 5′-GGGTTZTTGGG sequence results predominantly in cleavage at Z, with a lesser amount at the 5′-GGG and very little at the 3′-GGG. Based on our values of ΔG_{ht} for hole transport from G to Z and GGG, the relative equilibrium population in a ZTTGGG sequence should be ca. 70:1. Assuming similar rates for the chemical reactions leading to strand cleavage, cleavage at Z should predominate for this sequence. We note that Nakatani et al. [12] interpret their data differently; assuming that GGG forms a more stable hole trap than Z on the basis of gas phase calculations. This assumption is inconsistent with our data and that of Giese and Biland [20] and further requires that the rate constant for cleavage at Z be much faster than cleavage at GGG.

5.4 Theoretical Studies of Hole Transport Equilibria

The difference in stabilities of cation radicals located on G, GG, and GGG sequences was initially investigated by Sugiyama and Saito [14], who employed ab initio methods to calculate the gas phase ionization potentials of nucleobases stacked in B-DNA geometries. Their results indicated large differences in potential for holes on G vs GG (0.47 eV) and GGG (0.68 eV) sequences. A similar G vs GG difference was calculated by Prat et al. [62]. These values suggest that GG and GGG are, in fact, deep hole traps and they have been widely cited as evidence to that effect [54, 63].

Voityuk et al. [56] have employed semi-empirical calculations to determine the effects of neighboring bases on the relative energies of holes localized on a single nucleobase. Their calculations of the relative energies of 5′-XG$^{+\cdot}$Y triples provide relative energies of 0.0 eV for GG$^{+\cdot}$G and AG$^{+\cdot}$G triples and 0.13 eV for AG$^{+\cdot}$A and GG$^{+\cdot}$A triples. Thus a 3′-G lowers the potential by 0.13 eV, whereas a 5′-G has only a small effect. Our experimental value for hole transport equilibrium in 3GAGG is intermediate between the calculated values for AG$^{+\cdot}$G and GG$^{+\cdot}$A. Recently, Kurnikov et al. [64] have reported that the inclusion of solvation energies has a "leveling effect" on the calculated (ab initio) gas phase ionization potentials of G, GG, and GGG sequences. Solvation stabilization is largest for a hole localized on a single G and decreases with increasing delocalization. The net effect is a modest stabilization energy (<0.1 eV) for GG vs G and an even smaller stabilization energy for GGG. These authors point out that both their calculated stabilization energies and our experimental results for the trap depths for G and GG are within ~$2k_BT$ (where k_B is Boltzmann's constant).

It is interesting to note that our experimental values for the relative energies of G, GG, and GGG (0, 0.052, and 0.077 eV) are in excellent agreement with a Hückel analysis, according to which the stabilization energy for GGG should be larger than that for GG by $\sqrt{2}$. Conwell and Basko [65] have analyzed our results in terms of a delocalized polaron model, and both a small difference in A vs G oxidation potentials and a large decrease in interbase spacing (as much as 0.4 Å) are necessary to reproduce our data. Quantum calculations by Barnett et al.[66] also indicate extensive delocalization in a GAGG hole sequence.

Both our small experimental values for ΔG_{ht} and the values calculated by Voityuk et al. [56] and Kurikov et al. [64] indicate that GG and GGG form very shallow hole traps. This result is consistent with strand cleavage data for short duplexes containing a small number of G, GG, and GGG sequences (Table 2). It is also consistent with the analysis of strand cleavage data in longer duplexes containing multiple G and GGG sites, as reported by Giese and Spichty [58] and by Bixon and Jortner [38]. In view of these very small stabilization energies, it seems likely that holes are largely localized on a single G base rather than being strongly delocalized, as proposed in polaron models of charge transport in DNA.

6
Conclusions

Our studies reveal several important features of the dynamics of hole transport processes between a guanine and a secondary electron donor separated by one or two A:T base pairs. First, hole transport in the isoergic system GAG is slow (k_t~10^7 s^{-1}), relative to the rate constants for charge separation in an **Sa**-AG system (k_{cr}=2×10^{11} s^{-1}). These rates increase as the hole transport becomes more exergonic, in the series G<GG<GGG<Z. The markedly slower values of k_t vs k_{cs} are indicative of fundamental differences in these electron transfer processes. Second, rate constants for GAAGG or GAAZ systems are slower than those for GAG or GAZ systems by a factor of 10–20. A similar distance dependence is observed in the hole transport studies of Shavirovich et al. [43] and Kawai et al. [44] and in the strand cleavage studies of Meggers et al. [17]. Third, interstrand charge transport in a GA(CC) system is slower than intrastrand charge transport by only a factor of 7. A much large kinetic penalty is observed for charge transport in a GTGG sequence, which is 50–100 fold slower than in a GAGG sequence. This suggests the possibility that GT_nGG hole transport may occur via the A_n sequence in the complementary strand.

Our studies also provide the first experimental data for the free energy difference between different holes. The GAZ sequence has a value of ΔG_{ht}=−0.19 eV, somewhat smaller than the −0.3 eV difference in oxidation potentials of G vs Z. Much smaller values are obtained for GAGG and GAGGG sequences, for which ΔG_{ht}=0.052 and 0.077 eV, respectively. These values are smaller than those initially obtained from ab initio calculations [14], but are in agreement with more recent calculations which include the effects of solvation [64]. Our experimental values of ΔG_{ht} are similar in magnitude to $2k_BT$, suggesting that the holes may be localized on a single base. Combination of our results from equilibrium data with strand cleavage probabilities in short oligonucleotides provides relative rate constants of 1:0.2:0.1 for strand cleavage at G, GG, and GGG sites on a per-guanine basis.

The combination of shallow hole traps and slow hole transport is precisely the formula needed to explain the distance dependence of strand cleavage in longer oligonucleotides containing multiple G, GG, or GGG sites. Our kinetic data has been used by Giese [58] and Jortner [50] to model the results of strand cleavage studies. So long as hole transport is more rapid than the initial chemical events leading to strand cleavage, cleavage can be observed for dozens of base pairs from the locus of charge injection. We would, however, caution against the extrapolation of our data to longer-distance hole transport processes or to sequences other than those that we have investigated.

Have we pushed the use of time resolved transient absorption with kinetic modeling to its limit? We think not. Studies of the temperature dependence of hole transport in hairpins such as **3GAZ** should provide additional details about the mechanism of this processes. The use of donor-acceptor triplexes or dumbbells based on **3GASe** but possessing multi-donor sequences may permit us to unravel the complex dynamics of multiple hole transport processes. Strand cleavage studies in such systems would allow for the first direct correlation of dynamics and strand cleavage data for the same system. The use of time-resolved EPR may permit direct observation of guanine cation radicals and provide experimental information about the extent of charge delocalization. Finally, initial studies of electron injection to T or C via a $(G{:}G)_n$ bridge provide an entry to the study of electron migration processes in DNA [67].

Acknowledgments We thank our coworkers Jianqin Wu (synthesis), Ryan Hayes (transient spectroscopy), and Xiaobing Zuo (kinetic modeling) for their contributions to these studies and our colleagues Robert Letsinger for engaging our interest in DNA and Mark Ratner for discussions of hole transport processes. This research is supported by grants from the Division of Chemical Sciences, Office of Basic Energy Sciences, US Department of Energy, under contracts DE-FG02-96ER14604 (FDL) and by DE-FG02-99ER14999 (MRW).

References

1. Eley DD, Spivey DIT (1962) Trans Faraday Soc 58:411
2. Murphy CJ, Arkin MR, Jenkins Y, Ghatlia ND, Bossmann SH, Turro NJ, Barton JK (1993) Science 262:1025
3. O'Neil MP, Fieldew EM (1993) Adv Radiation Biol 17:53
4. Giese B (2000) Acc Chem Res 33:631
5. Schuster GB (2000) Acc Chem Res33:253
6. Treadway CR, Hill MG, Barton JK (2002) Chem Phys 281:409
7. Kovalsky OI, Panyutin IG, Budowsky EJ (1990) Photochem Photobiol 52:509
8. Saito I, Takayama M, Sugiyama H, Nakatani K (1995) J Am Chem Soc 117:6406
9. Muller JG, Hickerson RP, Perez RJ, Burrows CJ (1997) J Am Chem Soc 119:1501
10. Seidel CAM, Schulz A, Sauer MHM (1996) J Phys Chem 100:5541
11. Nakatani K, Fujisawa K, Dohno C, Nakamura T, Saito I (1998) Tetrahedron Lett 39:5995
12. Nakatani K, Dohno C, Saito I (2000) J Am Chem Soc 122:5893
13. Yoshioka Y, Kitagawa Y, Takano Y, Yamaguchi K, Nakamura T, Saito IK (1999) J Am Chem Soc 121:8712
14. Sugiyama H, Saito I (1996) J Am Chem Soc 118:7063
15. Hickerson RP, Prat F, Muller JG, Foote CS, Burrows CJ (1999) J Am Chem Soc 121:9423
16. Jortner J, Bixon M, Langenbacher T, Michel-Beyerle ME (1998) Proc Natl Acad Sci USA 95:12759
17. Meggers E, Michel-Beyerle ME, Giese B (1998) J Am Chem Soc 120:12950
18. Ly D, Sanii L, Schuster GB (1999) J Am Chem Soc 121:9400
19. Núñez ME, Hall DB, Barton JK (1999) Chem & Biol 6:85
20. Giese B, Biland A (2002) Chem Commun 2002:667
21. Lewis FD, Wu T, Zhang Y, Letsinger RL, Greenfield SR, Wasielewski MR (1997) Science 277:673
22. Lewis FD, Letsinger RL, Wasielewski MR (2001) Acc Chem Res 34:159
23. Lewis FD, Liu X, Liu J, Miller SE, Hayes RT, Wasielewski MR (2000) Nature 406:51
24. Letsinger RL, Wu T (1995) J Am Chem Soc 117:7323
25. Lewis FD, Liu X, Wu Y, Miller SE, Wasielewski MR, Letsinger RL, Sanishvili R, Joachimiak A, Tereshko V, Egli M (1999) J Am Chem Soc 121:9905
26. Weller A (1982) Zeit Phys Chem Neue Folg 133:93
27. Lewis FD, Wu T, Liu X, Letsinger RL, Greenfield SR, Miller SE, Wasielewski MR (2000) J Am Chem Soc 122:2889

28. Lewis FD, Wu Y (2001) J Photochem Photobiol C Rev 2:1
29. Bixon M, Jortner M (1999) Adv Chem Phys 106:35
30. Lewis FD, Liu J, Weigel W, Rettig W, Kurnikov IV, Beratan DN (2002) Proc Natl Acad Sci USA 99:12536
31. Lewis FD, Kalgutkar RS, Wu Y, Liu X, Liu J, Hayes RT, Wasielewski MR (2000) J Am Chem Soc 122:12346
32. Marcus RA (1965) J Phys Chem 43:679
33. Davis WB, Naydenova I, Haselberger R, Ogrodnik A, Giese B, Michel-Beyerle ME (2000) Angew Chem Int Ed 39:3649
34. Sistare MF, Codden SJ, Heimlich G, Thorp HH (2000) J Am Chem Soc 122:4742
35. Lewis FD, Liu X, Miller SE, Hayes RT, Wasielewski MR (2002) J Am Chem Soc 124:14020
36. Lewis FD, Liu J, Liu X, Zuo X, Hayes RT, Wasielewski MR (2002) Angew Chem Int Ed 41:1026
37. Kelley SO, Barton JK (1999) Science 283:375
38. Bixon M, Jortner J (2001) J Am Chem Soc 123:12556
39. Lewis FD, Liu X, Liu J, Hayes RT, Wasielewski MR (2000) J Am Chem Soc 122:12037
40. Lewis FD, Liu J, Zuo X, Hayes RT, Wasielewski MR (2003) J Am Chem Soc 125:4850
41. Lewis FD, Zuo X, Liu J, Hayes RT, Wasielewski MR (2002) J Am Chem Soc 124:4568
42. Lewis FD, Wu Y, Hayes RT, Wasielewski MR (2002) Angew Chem Int Ed 41:3485
43. Shafirovich VY, Dourandin A, Huang W, Luneva NP, Geacintov NE (2000) Phys Chem Chem Phys 2:4399
44. Kawai K, Takada T, Tojo S, Ichinose N, Majima T (2001) J Am Chem Soc 123:12688
45. Wagenknecht H-A, Rajski SR, Pascaly M, Stemp EDA, Barton JK (2001) J Am Chem Soc 123:4400
46. Bixon M, Giese B, Wessely S, Langenbacher T, Michel-Beyerle ME, Jortner J (1999) Proc Natl Acad Sci USA 96:11713
47. Rakhmanova SV, Conwell EM (2001) J Phys Chem B 105:2056
48. Zhang H-Y, Li X-Q, Han P, Yu XY, Yan Y-J (2002) J Chem Phys 117:4578
49. Berlin YA, Burin AL, Siebbeles LDA, Ratner MA (2001) J Phys Chem A 105:5666
50. Jortner J, Bixon M, Voityuk AA, Rösch N (2002) J Phys Chem A 106:7599
51. Bixon M, Jortner J (2002) Chem Phys 281:393
52. Berlin YA, Burin AL, Ratner MA (2002) Chem Phys 275:61
53. Berlin YA, Burin AL, Ratner MA (2000) J Phys Chem A 104:443
54. Troisi A, Orlandi G (2001) Chem Phys Lett 344:509
55. Olofsson J, Larsson S (2001) J Phys Chem B 105:10398
56. Voityuk AA, Jortner J, Bixon M, Rösch N (2000) Chem Phys Lett 324:430
57. Rak J, Voityuk AA, Marquez A, Rösch N (2002) J Phys Chem B 106:7919
58. Giese B, Spichty M (2000) Chem Phys Chem 1:195
59. Bixon M, Jortner J (2001) J Phys Chem A 105:10322
60. Hess S, Götz M, Davis WB, Michel-Beyerle ME (2001) J Am Chem Soc 123:10046
61. Giese B, Meggers E, Wessely S, Spormann M, Biland A (2000) Chimia 54:547
62. Prat F, Houk KN, Foote CS (1998) J Am Chem Soc 120:845
63. Berlin YA, Burin AL, Ratner MA (2001) J Am Chem Soc 123:260
64. Kurnikov IV, Tong GSM, Madrid M, Beratan DN (2002) J Phys Chem B 106:7
65. Conwell EM, Basko DM (2001) J Am Chem Soc 123:11441
66. Barnett RN, Cleveland CL, Joy A, Landman U, Schuster GB (2001) Science 294:567
67. Lewis FD, Liu X, Miller SE, Hayes RT, Wasielewski MR (2002) J Am Chem Soc 124:11280

Top Curr Chem (2004) 236:67–115
DOI 10.1007/b94412

DNA-Mediated Charge Transport Chemistry and Biology

Melanie A. O'Neill · Jacqueline K. Barton

Division of Chemistry and Chemical Engineering, California Institute of Technology, Pasadena, California 91125, USA
E-mail: jkbarton@caltech.edu

Abstract A range of photophysical and biochemical experiments have been conducted to establish DNA charge transport, to probe those parameters that influence DNA-mediated charge transport, and to explore the biological applications and consequences. Charge migration through the DNA base stack has been shown to result in oxidative damage 200 Å from the site of the remotely-bound oxidant, but this long-range reaction is exquisitely sensitive to perturbations in the intervening base stack. Spectroscopic measurements provide the timescale for the transfer between bases and highlight the dependence of coupling through the base stack on base dynamics. In exploiting the sensitivity of DNA charge transport to base pair stacking, DNA electrodes have been designed that detect, with high sensitivity, single base mismatches in DNA, as well as protein-dependent changes in DNA stacking. How proteins modulate long-range DNA charge transport has also been examined, and, in exploring charge transport within the biological milieu, oxidative damage to DNA within the cell nucleus is being probed.

Keywords DNA · Charge transport · DNA dynamics · Metallointercalators · DNA electrochemistry

1 Introduction

It was not long after the molecular structure of double helical DNA was revealed [1] that scientists began to ask whether inherent in the double helical stacked base pair structure were new functional characteristics for DNA [2]. Noting the resemblance of the array of aromatic bases in the DNA double helix to conductive, one-dimensional aromatic crystals, Eley and Spivey suggested that the DNA π-stack might provide a pathway for rapid charge separation [3]. This notion, that DNA can mediate motion of electronic charge, continues to fascinate and surprise us today. While a wealth of experimental evidence has now established, undeniably, that DNA can act as a conduit for rapid, and long-range charge transport (CT), the mechanisms and consequences of this role of DNA in diverse areas of chemistry and biology remain to be completely revealed.

Extended solid state π systems facilitate CT, particularly when doped [4–6]. The analogy between DNA and conductive solid state π-stacks therefore establishes that a requisite condition for CT may exist in DNA. DNA contains an array of heterocyclic aromatic base pairs, stacked at a distance of 3.4 Å, wrapped within a negatively charged sugar phosphate backbone [7] (Fig. 1). The interactions between the π electrons of the DNA base pairs provide the electronic coupling necessary for CT to occur.

Unlike solid state π-stacks, however, double helical DNA is a molecular structure. Here CT processes are considered in terms of electron or hole transfer and transport, rather than in terms of material conductivity. Moreover, the π-stack of DNA is constructed of four distinct bases and is therefore heterogeneous and generally non-periodic. This establishes differences in redox energetics and electronic coupling along the π-stack. The intimate association of DNA with the water and counterions of its environment further defines its structure and contributes to inhomogeneity along the mole-

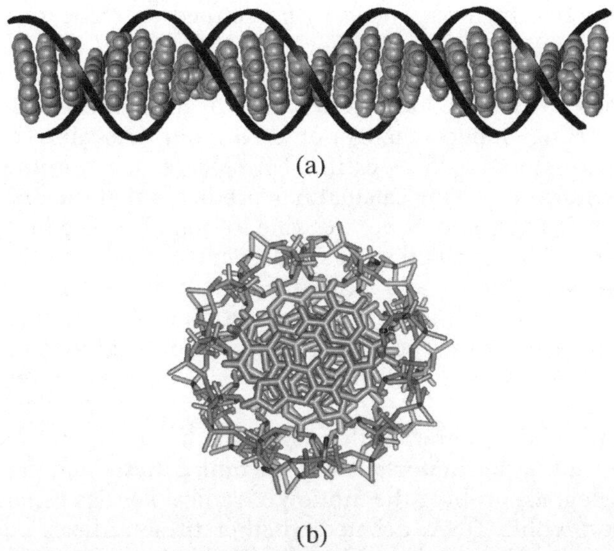

Fig. 1 The π-stack of double helical DNA. In this idealized model of B-DNA the stack of heterocyclic aromatic base pairs is distinctly visible within the sugar-phosphate backbone (schematized by ribbons): **a** view perpendicular to the helical axis; **b** view down the helical axis. It is the stacking of aromatic DNA bases, approximately 3.4 Å apart, that imparts the DNA with its unique ability to mediate charge transport. Base stacking interactions, and DNA charge transport, are exquisitely sensitive to the sequence-dependent structure and flexibility of DNA

cule. Perhaps most significantly, DNA is dynamic. Motions of the DNA bases in the picoseccond to millisecond time regimes [8–16], modulate stacking and coupling interactions between DNA bases on the time scale of charge transfer and transport reactions. In fact, what we have been learning, and describe in this Review, is that where the analogy between DNA and solid state π-stacked materials actually breaks down may be more interesting than where the analogy holds true.

The most straightforward test of the ability of DNA to transport charge is a direct measurement of DNA conductivity. Earlier experiments in this area generated vastly different results, and descriptions of DNA ranging from insulator to quantum wire emerged [3, 17–20]. The inherent complexities and variables of such measurements, particularly the nature of the electrical contacts and accessing the DNA π-stack, were likely sources of conflicting results. Despite the implementation of more sophisticated experimental tools, these issues remain controversial in more recent measurements of DNA conductivity [21–25]. Furthermore, the validity of extrapolating measurements made on dehydrated, structurally uncertain DNA samples exposed to high voltages to B-DNA molecules under physiological conditions is unclear. Overall, these studies reveal that methods designed to investigate current flow in conductive materials need to be refined in order to probe CT in DNA.

As with conductivity measurements, methods and results of theoretical treatments of CT in DNA have varied significantly. Mechanisms invoking hopping, tunneling, superexchange, or even band delocalization have been proposed to explain CT processes in DNA (please refer to other reviews in this text). Significantly, many calculations predicted that the distance dependence of CT in DNA should be comparable to that observed in the σ-systems of proteins [26]. This prediction has not been realized experimentally. The dichotomy between theory and experiment may be related to the fact that many early studies gave insufficient consideration to the unique properties of the DNA molecule. Consequently, CT models derived for typical conductors, or even those based on other biomolecules such as proteins, were not adequate for DNA.

Charge and radical migration through DNA has profound biological implications, including an important role in mutagenesis and carcinogenesis. Radiation biologists, probing the motion of radical species generated by ionizing radiation within DNA, debated whether these radicals could traverse two or two hundred base pairs within the double helix [27]. Here the diversity of experimental conditions and methods for accurately assessing the distance of migrating radicals were key issues.

Chemists approach DNA CT on a molecular level by examining charge transfer and transport reactions between electron donors and acceptors mediated by and/or directly involving DNA. These experiments aim to measure the rate constants and yields of charge transfer reactions as a function of distance. Our initial investigations of DNA CT exploited the photophysical and chemical properties of polypyridyl metal complexes. First experiments revealed enhanced efficiency of photoinduced electron transfer between complexes when weakly bound to DNA; the rate constant for luminescence quenching of $Ru(phen)_3^{2+}$ (phen=9,10-phenanthroline) by tris(phenanthroline) complexes of Co(III) and Rh(III) increased by two orders of magnitude [28, 29]. The role of the DNA π-stack in these charge transfer reactions was immediately revealed by our metallointercalator redox probes, that bind DNA avidly by intercalation between the bases [30–32]. Oxidative quenching of dipyridophenazine (dppz) complexes of Ru(II) by phenanthrequinone diimine (phi) complexes of Rh(III) was found to be extremely rapid ($k>10^{10}$ s^{-1}), only when both complexes were intercalated into DNA base stack. Transient absorption measurements of DNA-mediated quenching of dppz complexes of Ru(II) or Os(II) by $Rh(phi)_2bpy^{3+}$ (bpy=2,2′-bipyridine) facilitated direct observation of the oxidized Ru(III) or Os(III) intermediate, confirming electron transfer (ET) as the quenching mechanism [33, 34].

These preliminary studies established that charge transfer reactions did indeed proceed in DNA, and suggested that these reactions may occur rapidly over a long distance, and thereby be mechanistically distinct from ET in proteins. Particularly provocative was the description of DNA-mediated CT as largely insensitive to distance, but strongly dependent on stacking interactions between the redox reagents and the DNA base pairs. Not surprisingly, however, alternative explanations were proposed [35, 36], and using different DNA-bound donors and acceptors, CT behavior reminiscent of proteins was observed [37, 38], suggesting that no new paradigm was required to describe DNA CT. This controversy, however, most likely originated in the lack of control over and characterization of the DNA assemblies, particularly in terms of the CT distance and association of the redox probes with the DNA.

Being intrigued by the potential of this CT chemistry, we devised new tools to accurately probe and describe it. In particular, methods for precise control and characterization of our donor-DNA-acceptor assemblies were developed. Structurally well-characterized DNA assemblies containing covalently bound and stacked donors and acceptors were prepared. We also expanded our repertoire of redox active probes to include an array of metallointercalators, organic intercalators, and DNA base analogues. As first investigations hinted that reactions over varied distance and time regimes may be characteristic of DNA CT, we adapted a versatile combination of spectroscopic, biochemical and electrochemical methods to probe DNA CT under diverse conditions. Here we highlight research in our laboratory to probe this chemistry using these tools and assemblies. Results emanating from other laboratories are described in other chapters of this text. Throughout, we wish to emphasize what we have learned from these varied experiments: that the structure and dynamics of the DNA π-stack make charge transfer and transport possible, and likewise regulate its rate constants, efficiencies and distance dependence. Our current questions regarding DNA CT are undoubtedly linked to how the unique structure and dynamics of DNA determine chemical and biological functions for CT.

2 Experimental Approaches to Studies of DNA-Mediated Charge Transport

Our approach to studies of CT in DNA relies on two key features. First is the use of well-characterized DNA assemblies, which include redox probes that are strongly coupled to the DNA π-stack. The importance of well-characterized DNA assemblies, including the redox participants and DNA bases, cannot be overstated. Differences in structural and energetic properties of DNA assemblies, particularly when unaccounted for, may be responsible for drastically different conclusions regarding DNA CT. Furthermore, in order to characterize the DNA π-stack as a medium for CT, it is necessary to employ redox probes that are directly coupled to the π-stack.

Second is the application of a wide range of experimental designs and techniques. DNA CT is observed in a diverse array of systems over different distance and time regimes. Consequently, a versatile approach which draws upon complementary methods is required to explore different facets of this chemistry and develop a complete picture. We interrogate a variety of nucleic acid assemblies using spectroscopic, biochemical and electrochemical tools to define mechanistic features, exploit biological applications, and explore biological consequences of DNA CT.

2.1 Metallointercalators, Organic Intercalators and Modified Bases as Probes

Redox participants are chosen to facilitate spectroscopic, biochemical and electrochemical probing of DNA CT. These include metallointercalators, organic intercalators, and modified bases that possess useful, well-described, and varied redox, photophysical and photochemical properties (Table 1). Our probes are readily incorporated into DNA assemblies where CT distances ranging from 3.4 to 200 Å and driving forces spanning over two volts can be modulated with certainty. Most importantly, all redox probes which afford fast and/or efficient CT through DNA are well-coupled to the π-stack.

Table 1 Probes of DNA-mediated CT and their redox potentials (in solution) versus NHE

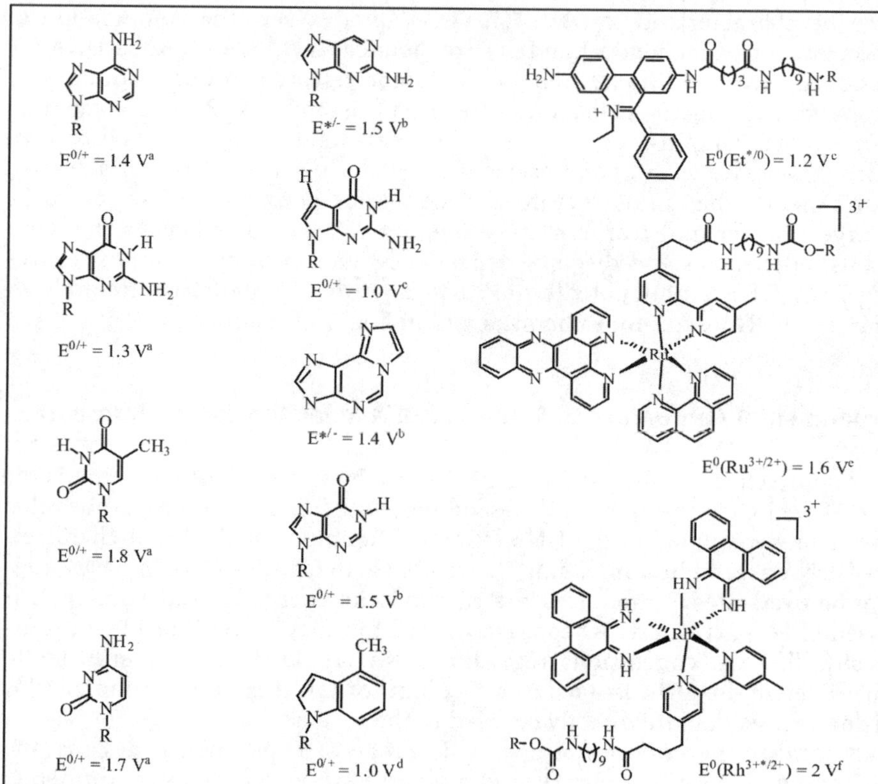

[a][39], [b][40], [c][41], [d][42], [e][30], [f][43]

2.1.1
Metallointercalators

A number of metallointercalators are commonly employed to study long-range CT in DNA, notably phenanthrequinone diimine (phi) complexes of Rh(III) (for example $[Rh(phi)_2bpy']^{3+}$), and dipyridophenazine (dppz) complexes of Ru(II) (for instance $[Ru(phen)(bpy')(dppz)]^{2+}$) [44]. These ligands facilitate tight binding ($K \geq 10^6$ M^{-1}) of the octahedral Rh(III) and Ru(II) complexes to DNA by intercalation, and the interactions of these metallointercalators with DNA have been extensively characterized [45]. In addition, we have devised facile synthetic schemes for covalent attachment of these complexes to DNA [46]. Due to their positive charge, ET reactions involving these metallointercalators are charge shift reactions.

Phi complexes of rhodium bind DNA avidly by intercalation of the phi ligand [47, 48]. High resolution NMR studies indicated that these complexes bind from the major groove with a sequence specificity that is modulated by the ancillary, non-intercalating ligands [49, 50]. This binding mode was recently confirmed by a 1.2 Å crystal structure of the Δ-α-[Rh(R,R-dimethyltrien)phi]$^{3+}$ complex intercalated into a DNA octamer [51]. In addition, five independent views of the complex established that intercalation induces minimal perturbation of the DNA π-stack, either globally or locally (Fig. 2). No kinking or bending is associated with intercalation. The inserted phi ligand can essentially be described as an additional base pair.

The photochemistry of phi complexes of rhodium makes them particularly amenable probes for DNA CT [52]. Exciting these complexes with ultraviolet light (λ=313 nm) results in direct scission of the DNA sugar-phosphate backbone via hydrogen abstraction from the sugar near the photoexcited intercalated phi ligand. Consequently, the binding site of the complex is marked. Exciting these complexes with visible light ($\lambda \geq 365$ nm) generates a potent photooxidant ($E^0(Rh^{3+*/2+})$~2 V vs NHE) [31] which, when intercalated into DNA, leads to oxidative damage at guanines. The oxidation potential of these photoexcited phi complexes should be sufficient to oxidize all four DNA bases as well as thymine dimers (Table 1).

Dipyridophenazine complexes of Ru(II) have been dubbed "molecular light switches" due to their remarkable photophysical properties when bound to DNA [53, 54]. Although these dppz complexes display luminescence in organic solvents, their luminescence is quenched in aqueous solution, as a result of proton transfer from water to the nitrogen atoms of the phenazine [55]. Intercalation into the DNA π-stack protects the phenazine nitrogen atoms from water, and, consequently these molecules light up when bound to DNA. This interesting photophysical feature can be exploited, for instance, to delineate the DNA binding mode of the dppz ligand. NMR investigations of partially deuterated Δ-$[Ru(phen)_2dppz]^{2+}$ bound to a DNA hexamer indicated that the dppz ligand intercalates from the major groove with two distinct binding modes [56]. In the symmetrical orientation, the phenazine nitrogens are protected from solvent, whereas in the asymmetrical ori-

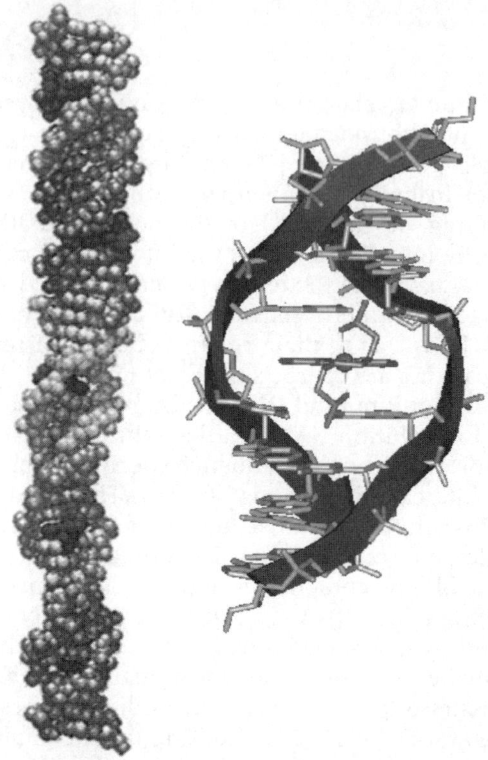

(a) (b)

Fig. 2 A 1.4 Å crystal structure reveals that intercalating phi ligands do not disrupt the DNA π-stack. Shown are: **a** the five Δ-α-[Rh(R,R-dimethyltrien)phi]$^{3+}$-DNA octamers stacked end-to-end in the asymmetric unit of the crystal, and **b** a view from the major groove of Δ-α-[Rh(R,R-dimethyltrien)phi]$^{3+}$ intercalated in the DNA duplex, 5′-G-dIU-TGCAAC-3′. The intercalator is inserted as an additional DNA base step with minimal perturbation of DNA π-stack. Adapted from [51]

entation, the phenazine nitrogens are more solvent exposed. Thermodynamic studies reveal that binding to DNA is avid [57].

Photoexcitation of Ru(II) dppz complexes generates a metal-to-ligand charge transfer excited state that is localized on the dppz ligand. This is a particularly salient feature for investigations of DNA-mediated CT, because excitation of the metallointercalator directs the charge transfer into the π-stack rather than on an ancillary ligand [58]. We exploit the redox properties of photoexcited Ru(II) ($E^0(Ru^{2+*/3+})$~0.6 V vs NHE) [30, 34] to generate a potent ground state Ru(III) oxidant ($E^0(Ru^{3+/2+})$~1.6 V vs NHE) [34] in situ by the flash quench technique [59] (Fig. 3). Here, the excited state complex, Ru(II)*, undergoes an ET reaction with a non-intercalated oxidative quencher (for example methyl viologen, $Ru(NH_3)_6$) yielding the ground state Ru(III) oxidant intercalated within the DNA π-stack. This Ru(III) complex is

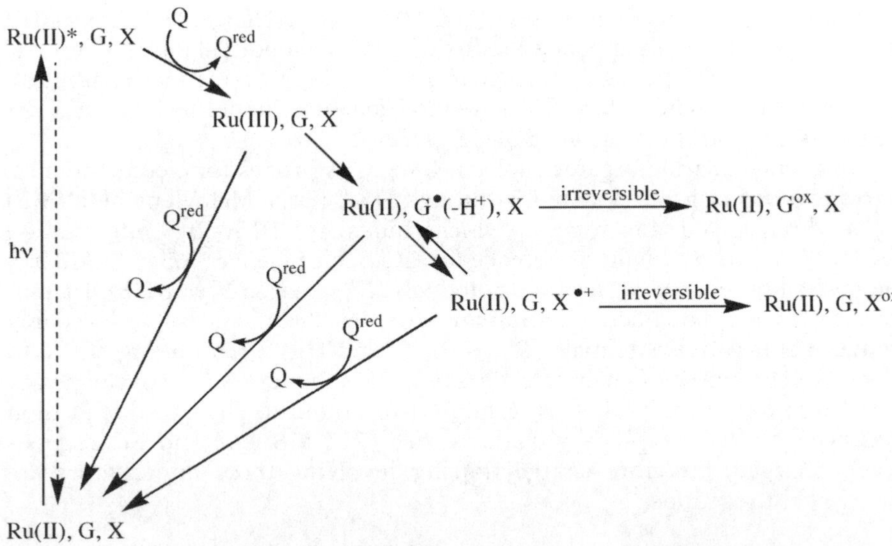

Fig. 3 Flash quench scheme for in situ generation of Ru(III). Oxidative quenching of a photoexcited Ru(II) intercalator by a diffusible quencher (Q) generates Ru(III), a potent ground state oxidant which can react with guanine bases in DNA. Once oxidized, the guanine radical cation can rapid deprotonates ($>10^7$ s^{-1}). The resultant guanine radical may be irreversibly trapped leading to permanent DNA lesions. Conversely hole migration to additional traps, X, positioned elsewhere in the DNA duplex may occur. Competing with these processes is back reaction with the reduced quencher to regenerate ground state Ru(III)

sufficiently potent to oxidize guanine, and perhaps adenine, based on the potential of these nucleosides as measured in solution (Table 1). The flash quench technique, originally designed for exploring ET in proteins [60], has proven invaluable for the generation of powerful ground state oxidants bound to DNA. A significant feature of the flash quench method is that a diffusion-controlled reaction is required to generate the Ru(III) oxidant. This sets the time-scale for the direct measurement of reaction kinetics (typically $k\sim\leq 10^7-10^8$ s^{-1}).

2.1.2
Organic Intercalators

Intercalating molecules, such as the classic organic intercalator, ethidium (Et), have likewise been significant in our efforts to understand DNA-mediated CT (Table 1). The intercalative interactions and fluorescence properties of Et in the presence and absence of DNA have been extensively characterized and are highly amenable to investigations of DNA CT. Ethidium intercalates tightly into DNA ($K\sim10^6$ M^{-1}) [61, 62], and its luminescence intensity is significantly enhanced upon DNA binding [63]. The excited state of Et formed upon irradiation with visible light, has a potential that is insufficient to oxidize the natural DNA bases (E^0(Et$^{*/0}$)=1.2 V vs NHE) [41]. This allows selective reaction with appended redox reagents or modified bases of low oxidation potential such as 7-deazaguanine (^{Z}G). Moreover, functionalization of Et with methylene tethers facilitates covalent attachment to DNA in order to precisely control the distance of CT [41, 64].

Other intercalating organic molecules serve as probes for electrochemical investigations of ground state CT through DNA films. Methylene blue (MB) is a three-ringed heterocycle which binds to DNA by intercalation (K~10^6 M^{-1}) with a slight preference for GC-rich sequences [65–68]. MB can be reversibly reduced at negative potentials (E^0_{red}~−0.25 V vs NHE). Daunomycin (DM), a redox active antitumor agent [69], can likewise be reversibly reduced at negative potentials (E^0_{red}~−0.4 V vs NHE). Crosslinking of DM to the exocyclic amine of guanine generates a DNA duplex with a covalently appended redox probe [70]. Furthermore, DM bound to DNA in this fashion has been crystallographically characterized [71]. MB, DM and Et are positively charged; therefore charge transfer involving these intercalators are charge shift reactions.

2.1.3
Modified Bases

We use modified DNA bases to tune the photophysics and redox properties of bases in DNA with minimal structural impact (Table 1). DNA oligonucleotides containing base analogues at precisely defined positions are readily prepared by standard solid phase synthesis and easily characterized. The hydrogen bonding and stacking interactions of base analogs within DNA duplexes are often comparable to the natural bases, and may be characterized by a variety of spectroscopic tools. Crystallographic or NMR structures of many DNA duplexes possessing base analogues now exist. The base analogues we employ are neutral molecules and their CT reactions with natural DNA bases involve charge separation.

The fluorescent base analogues, 1-N^6-etheno-adenine (εA) and 2-aminopurine (Ap) are two examples that illustrate the utility of modified bases as probes. Although the natural DNA bases are essentially non-fluorescent [72–73], both adenine analogues, εA [74–76] and Ap [77, 78], emit strongly in solution and in DNA. These probes can be selectively excited in DNA to generate powerful photooxidants (Ap: $E^{0(*/-)}$~1.5 V; εA: $E^{0(*/-)}$~1.4 V vs NHE) [40] which can initiate CT chemistry within the base stack. Equally important, the photoexcited molecules, particularly Ap*, are remarkably sensitive to their environment. Consequently, the fluorescence of Ap* acts as a detailed reporter of structure and dynamics within the double helix [8, 9, 79–84]. The solution structures of DNA duplexes containing εA [85] and Ap [8] have both been characterized by NMR. Opposite thymine, Ap is base paired and stacked in duplex DNA in a manner which is very similar to adenine. Conversely, εA assumes a non-planar conformation and exhibits correspondingly poorer stacking interactions with neighboring bases.

Base analogs are often chosen in order to access different redox potentials within DNA. Typically these molecules differ very little from the natural DNA bases such that the change in redox potential dominates over any minor structural perturbation. Examples are 7-deazaguanine (^{z}G) and 7-deazaadenine (^{z}A) which are derived by replacing the N-7 atom of guanine and adenine, respectively, by a C–H. This trivial structural modification induces a dramatic change in oxidation potential of ~300 to 400 mV [41, 86, 87]. Inosine (I) is another guanine analog with particularly useful redox properties. Replacement of the exocyclic amine of guanine by a hydrogen induces a 200 mV increase in oxidation potential [40]. Substitution of I for G in a DNA duplex results in the loss of one hydrogen bond in the base pair with cytosine. Other base analogues exist which form no hydrogen bonds with natural bases, but instead are stabilized by base stacking interactions in duplex DNA [88]. An example with particularly desirable features for our CT investigations is methyl indole (M). This molecule has an exceptionally low redox potential ($E^{0(0/+)}$~1 V vs NHE) [42] and can therefore serve as a relatively deep hole trap. In addition, since the methyl indole radical cation exhibits significant absorption in the visible region [89], methyl indole is an ideal probe for investigations of CT by transient absorption spectroscopy.

2.2
Spectroscopic, Biochemical and Electrochemical Approaches

Using a variety of spectroscopic techniques, we have observed the time-resolved dynamics of charge transfer and transport on time-scales from femtoseconds to milliseconds. Our time-resolved studies are coupled to steady-state spectroscopy in order to evaluate both reaction rate constants and yields. Fluorescence spectroscopy is employed to measure excited state lifetimes of redox probes within the DNA π-stack, and to monitor charge transfer quenching of these photoexcited reagents. This defines the time-scale and efficiency of the reaction. Transient absorption spectroscopy is employed to directly observe the transient intermediates along the CT pathway, and to monitor back electron transfer (BET). In all spectroscopic investigations, we exploit the photon energy to initiate charge transfer between our photoexcited donor and an acceptor positioned elsewhere in the DNA helix. Importantly, the information obtained from each spectroscopic method is governed by the time-scale accessible to the instrument. Consequently, each experiment represents a discrete piece of the overall CT dynamics which encompass a much wider time scale.

Biochemical assays truly probe "chemistry at distance" by examining DNA damage and repair products generated by long-range DNA-mediated CT. The experiments typically involve steady-state irradiation of an intercalated photooxidant tethered to DNA, leading to the injection and migration of a hole through the duplex. These holes are ultimately trapped at the sites of lowest oxidation potential, which are guanine sites in native DNA. Often multiple guanines, particularly guanine doublets, are used as hole traps. Selective oxidation at the 5′-G of guanine doublets has become a hallmark of

CT-induced damage, consistent with theoretical predictions that the HOMO of guanine doublets is localized on the 5′-G [90]. Conversely, non-specific oxidation of both guanines is indicative of an alternative chemistry, for instance reaction with singlet oxygen.

The oxidized guanines subsequently react with water and/or oxygen to yield permanent lesions that are revealed as strand breaks upon enzymatic treatment or reaction with base (such as piperidine) [91]. The resultant fragments are separated, visualized and quantified by polyacrylamide gel electrophoresis (PAGE) or high performance liquid chromatography (HPLC). What renders biochemical assays so valuable is that they are not limited by time, and can therefore examine the final product of a CT reaction that has occurred over considerable distance. However this same feature is responsible for the limitation of biochemical methods; the lack of real-time observation precludes the detection of intermediates and the possibility to distinguish between charge injection, migration and trapping.

We have also developed novel methods for electrochemical probing of DNA-mediated CT [92, 93]. Here we exploit molecular self-assembly to generate well-defined monolayers of thiol-modified DNA duplexes on gold electrodes. A redox active intercalator, such as MB or DM, is bound to the DNA at a distance from the gold surface. This intercalator acts as a reporter of electrochemically-initiated CT through the intervening DNA bridge; reduction of the distantly bound intercalator is monitored by, for instance, cyclic voltammetry or chronocoulometry. The yield of reduced intercalator therefore provides a measure of the efficiency of CT through DNA. In contrast to other techniques, the electrochemical methods probe electron, rather than hole, transport (HT) between ground state, rather than photoexcited, molecules. Correspondingly, the distance, time and energetic regimes are distinct. Notably the charge must traverse the alkane-thiol linker before accessing the DNA π-stack. As this process is considerably slower [94] than measured rate constants for CT in DNA, it may not be possible to obtain direct information on the rates of CT through these DNA films.

3 Probing DNA Charge Transfer and Transport over Different Distance, Time and Energetic Regimes

Preliminary investigations of DNA CT suggested that it may be quite distinct from CT through proteins, in distance, time, and energetic regimes. In addition these studies hinted that a strongly coupled π-stack including the redox participants and the intervening DNA bases is essential to CT. These features provide the impetus and focus for our continued exploration of DNA-mediated CT.

3.1
Spectroscopic Investigations of Charge Transport Through DNA

3.1.1
Charge Transport Between DNA Intercalators

The first experiments characterizing DNA-mediated CT over a precisely defined distance between covalently appended redox probes were reported in 1993 [95]. Remarkably, the luminescence of a photoexcited Ru(II) intercalator was quenched by a Rh(III) intercalator fixed to the other end of a 15-mer DNA duplex over 40 Å away (Fig. 4). Furthermore, non-intercalating, tethered Ru(II) and Rh(III) complexes did not undergo this quenching reaction. In this way the importance of intercalative stacking for efficient CT was demonstrated.

To further probe this CT chemistry, in particular the distance dependence and role of π-stacking interactions, we considered reactions between photoexcited Et and $[Rh(phi)_2bpy']^{3+}$ covalently appended to DNA and intercalated within the base stack. [64] (Fig. 5). Here, as reported for the metallointercalators, charge transfer quenching of Et^* luminescence by the Rh(III) intercalator was observed over distances of 20–30 Å. Significantly, although a shallow distance dependence was observed in the yield of CT, the fluorescence decay rate of the Et-modified assemblies was unaffected by the Rh quencher. From these observations we proposed that the rate constant for charge transfer was fast compared to the time scale of our instrumentation (10^{-10} s) and exhibited no significant variation with distance; what we observed was the subpopulation of molecules that were not quenched, presumably owing to a defect in the intervening stack. This was our first suggestion that the distance dependence of CT yield might reflect the increased probability of base destacking with increasing donor-acceptor distance, rather than a decrease in CT rate constant.

To test this sensitivity of photoinduced quenching to perturbations in the intervening base pair stack, an assembly was prepared containing an intervening CA base mismatch which disrupts locally the π-stack between the donor and acceptor. We observed that the yield of CT was reduced in the DNA

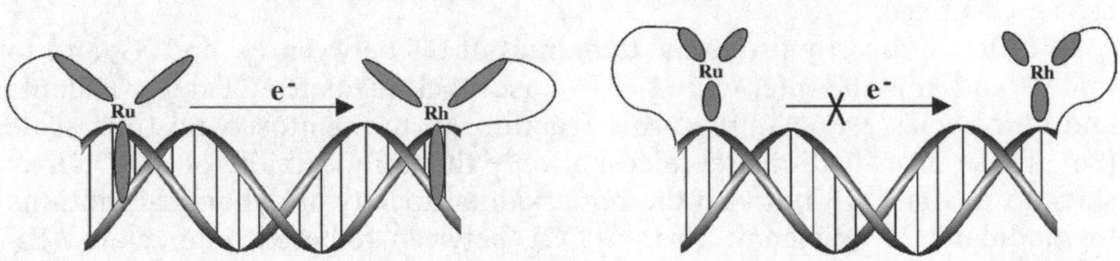

Fig. 4 Coupling of the redox participants to the DNA π-stack is requisite to DNA-mediated charge transport. Rapid ($>10^9$ s^{-1}) photoinduced electron transfer occurs between the metallointercalators, $[Ru(phen)_2dppz]^{2+}$ and $[Rh(phi)_2phen]^{3+}$, when they are tethered to opposite ends of a DNA duplex over 40 Å apart. Conversely, electron transfer does not occur between non-intercalated Ru(II) and Rh(III) complexes tethered to DNA

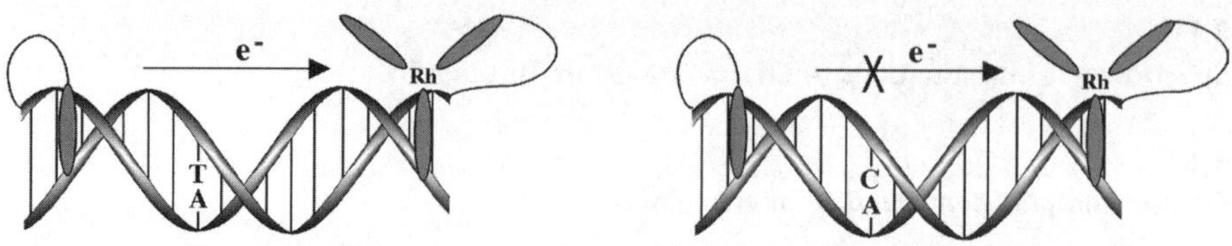

Fig. 5 Charge transfer occurs through the DNA π-stack and is strongly dependent on minor base stack perturbations. Photoinduced electron transfer is observed from tethered intercalated ethidium to a rhodium intercalator bound to DNA up to 30 Å away. The efficiency of electron transfer is drastically reduced in the presence of a stacking disruption, here a single base-pair mismatch

possessing the mismatch (Fig. 5). This confirmed first that the pathway of CT is through the DNA base stack not the sugar-phosphate backbone. Furthermore this result emphasizes the sensitivity of DNA CT to perturbations in this base stack. Unlike CT in proteins, here CT depends not only on distance but sensitively upon the intervening π-stacked structure.

3.1.2
Charge Transport Between an Intercalator and a DNA Base

In order to examine long range CT reactions in which DNA bases are participants, we prepared DNA duplexes containing ^{Z}G and covalently tethered Et (Fig. 6). As ^{Z}G has an oxidation potential which is ~300 mV below guanine, selective quenching of photoexcited Et by ^{Z}G is observed when Et is intercalated within mixed sequence DNA. Using steady-state fluorescence and nanosecond time-correlated single photon counting (TCSPC), we characterized the CT reactions between Et* and ^{Z}G as a function of distance, DNA sequence, and base stacking [41]. The CT was found to proceed on a subnanosecond time-scale over distances of 6 to 27 Å. Furthermore, while the yield of CT exhibited a shallow distance dependence, both the yield and distance dependence were dramatically influenced by subtle changes in base sequence, including intervening mismatches. These data were again consistent with ultrafast CT sensitively modulated by the nature of the intervening DNA π-stack, even in assemblies where the distance between donor and acceptor was fixed.

In order to directly probe the dynamics of CT between Et* and ^{Z}G, and to understand *how* the intervening DNA base stack regulates CT rate constants and efficiencies, we examined this reaction on the femtosecond time scale [96]. These investigations revealed not only the unique ability of the DNA π-stack to mediate CT, but also the remarkable capacity of dynamical motions to modulate CT efficiency. Ultrafast CT between tethered, intercalated Et* and ^{Z}G was observed with two time constants, 5 and 75 ps, both of which were essentially independent of distance over the 10–17 Å examined. Significantly, both time constants correspond to CT reactions, as these fast decay components were not detected in analogous duplexes where the ^{Z}G was re-

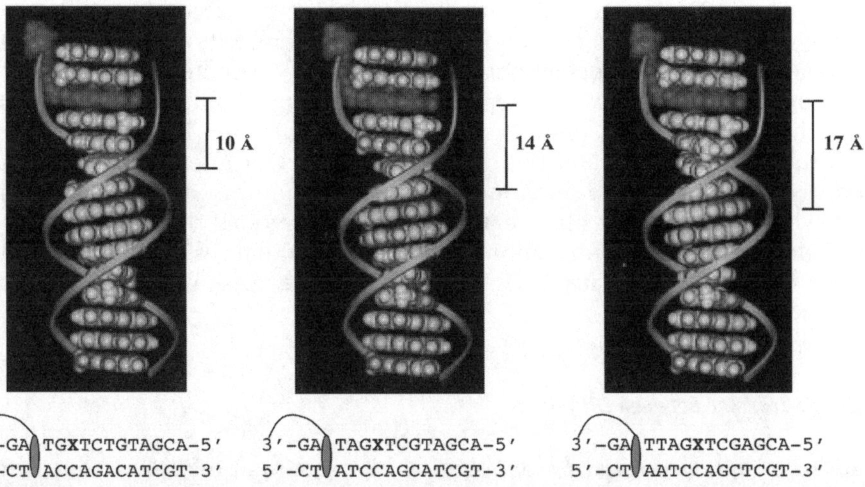

Fig. 6 Dynamic molecular motions can gate DNA-mediated charge transport. Two time constants (5 and 75 ps) are observed for hole transfer from photoexcited ethidium, tethered and intercalated near the end of a 14-base pair DNA duplex, to a base analog, 7-deazaguanine, in DNA. The 5 ps time constant, which is independent of distance between 10–17 Å, is due to direct hole transfer, while the 75 ps time constant corresponds to reorientation of the ethidium before hole transfer. Adapted from [96]

placed by G (where the oxidation reaction is not favorable). We therefor assigned the 5 ps component as the inherent rate for direct CT between Et^* and ^zG. The longer, 75 ps decay was attributed to reorientational motion of Et^* within its binding site, prior to CT. This motion positions the Et^* in the correct conformation for CT. Reorientation of Et^* is slow relative to the 5 ps required for CT between Et^* and ^zG when in CT-favorable conformation. This assignment was supported by fluorescence anisotropy measurements which established the 75 ps time component as the time for motion of intercalated Et^*. Also consistent with this proposal are investigations of the dynamics of ET between Et^* and 7-deaza-2′-deoxyguanosine triphosphate (dZTP) in solution [97]. Here the relative orientational motions within Et^*-dZTP complexes were found to be the rate determining step for ET.

Our investigations of ultrafast CT between Et^* and ^zG suggest that DNA-mediated CT may be gated by molecular motions within the DNA assembly, in this case motions of Et^*, the photoexcited hole donor. Conformational gating has been observed for ET reactions through other molecular assemblies [98], and is certainly expected to be significant to DNA CT, given the timescale of molecular motions within DNA. Indeed, the role of conformational gating in CT through macromolecules, including DNA, has recently been addressed theoretically [99, 100].

Stacking dynamics and conformational gating also provide an explanation for the distance dependence of the yield of CT between Et^* and ^zG, a dependence which was not manifested in the CT rates [96]. Dynamic motions of the DNA bases and redox probes on the picosecond to millisecond time scale generate a distribution of conformations within the DNA assembly. Only certain conformations are expected to facilitate CT. For these favorable conformations, sufficient donor-acceptor coupling is achieved during the lifetime of the photoexcited donor, in this case Et*, and CT occurs with a characteristic rate. Increasing the number of intervening base pairs between the donor and acceptor increases the number of molecules which must be appropriately aligned for CT. Correspondingly, the probability of forming CT-active conformations decreases. This reduces the number of discrete CT events, and causes the yield of CT to decrease with increasing donor-acceptor distance.

3.1.3
Charge Transfer Between DNA Bases

Spectroscopic investigations employing intercalating probes have uncovered ultrafast DNA-mediated CT that is regulated by static and dynamic stacking interactions. To further elucidate the role of the DNA bases in this chemistry, we have been exploring systems where DNA bases act as both the redox participants and the intervening bridge. Here charge transfer between DNA bases and the influence of stacking interactions among the DNA bases can be explicitly probed. In this effort we have exploited non-natural DNA bases, in particular the fluorescent base analogues εA and Ap.

We have conducted a systematic study of DNA-mediated oxidation of G and ^{z}G by either Ap or εA within 12-mer DNA duplexes over donor-acceptor separations of 3.4 to 13.6 Å [40, 101]. This investigation combined steady-state fluorescence with nanosecond and femtosecond fluorescence and transient absorption spectroscopy to generate an accurate picture of base-base CT reactions. While the reactivities of εA and Ap towards CT with G or ^{z}G are quite similar in solution, striking differences were observed for the same CT reactions within duplex DNA. Charge transfer involving Ap* is rapid ($k \sim 10^{10}$–10^{11} s^{-1}) and characterized by a relatively shallow decrease in both the yield and the rate constant with distance. Conversely, charge transfer involving εA* is several orders of magnitude slower and exhibits a steep distance dependence. High-resolution NMR investigations of duplexes containing εA [85] and Ap [8] reveal important differences in the structure of these duplexes, namely in stacking within the helix. The sterically bulky εA does not pair with T and exists in a non-rigid conformation that is not effectively stacked with the DNA bases. Ap is base paired and stacked in a manner which is very similar to the native DNA bases. Here the dramatic differences in charge transfer rate constant and efficiency can be directly correlated to differences in DNA base stacking.

In many investigations of CT, pendant redox probes interact with both bases of a base pair. However, studies of base-base charge transfer can differentiate between discrete intra- and interstrand reactions (Fig. 7). These investigations further attest to the critical role of base stacking in DNA-mediated CT. In B-DNA duplexes, stacking interactions are largely restricted to

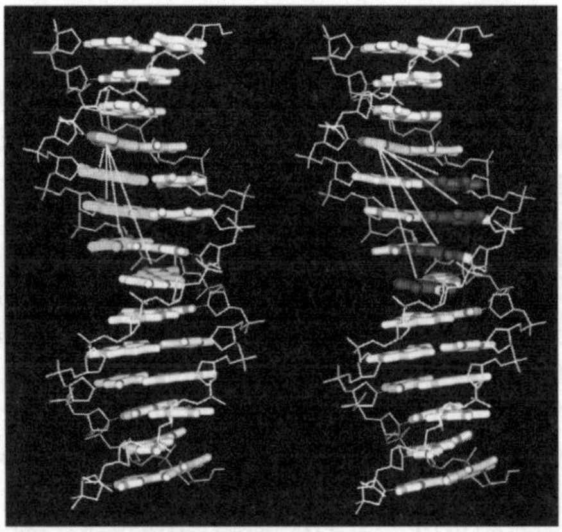

Fig. 7 Intra- and interstrand charge transfer in DNA are distinct. Experiments monitoring charge transfer between photoexcited 2-aminopurine, a base analog, and guanine or 7-deazaguanine established that intrastrand charge transfer is faster, more efficient, and proceeds with a different distance dependence than interstrand charge transfer. In DNA, intra- and interstrand pathways are characterized by different base stacking interactions and electronic coupling. Adapted from [40]

bases within the same strand [7]. This can be rationalized, qualitatively, by the lack of significant overlap between bases on opposite strands. The rate constants and efficiencies of base-base CT parallel this stacking arrangement. Therefore, intrastrand charge transfer between Ap* and G [101] was observed to be $\sim 10^3$ times faster than the analogous interstrand reaction [40]. A kinetic penalty for interstrand charge transfer has also been observed for HT between G bases [102], and between Ap radical cation and G [103]. Due to the lack of significant interstrand stacking in B-DNA duplexes, the interstrand reaction presumably requires charge transfer across a hydrogen bond of a base pair. These experiments highlight a fundamental feature of DNA charge transfer and transport; direct coupling of reactants through stacking is requisite for fast reaction kinetics.

A dramatic demonstration of the sensitivity of CT to stacking emerged from our investigation of the influence of DNA directional asymmetry on HT between DNA bases [104]. Remarkably, for reactions through the same DNA bridge, over the same distance, and with the same driving force, HT from Ap* to G in the 5′ to 3′ direction is more efficient and less dependent on distance than HT from 3′ to 5′. These differences in HT efficiency reflect variations in base-base coupling within the DNA assemblies. Undoubtedly base-base coupling is a critical parameter in DNA CT, and this base-base coupling is strongly dependent on subtle structural nuances of the DNA double helix.

More recently we have turned our attention to the influence of dynamic variations in stacking on base-base CT. Here again Ap* is particularly useful given its demonstrated charge transfer chemistry in DNA, and its ability to report on the structure and dynamics of the DNA environment. Using specific Ap-containing DNA duplexes and analogous DNA:RNA hybrids we characteristically tuned the base pair dynamics within the charge transfer bridge [105]. Fluorescence investigations correlated the distance dependence of charge transfer yield with the conformational flexibility of the bridge; a more shallow distance dependence was evident in duplexes with increased base pair dynamics. These results are consistent with the notion of CT via specific, well-coupled conformations of the redox reagents and DNA bridge. Within a certain regime, dynamic motion of the DNA base pairs provides access to and sampling of these conformations. Therefore, the number of discrete charge transfer events may be larger for duplexes with increased conformational flexibility. In such cases, the distance dependence of charge transfer yield will fall off more gradually when compared to duplexes which have more restricted access to these conformations.

Studies of base-base CT in B-form DNA duplexes versus A-form DNA:RNA hybrids also confirmed our fundamental tenet that the pathway of efficient charge transfer will be the well-stacked pathway. In A-form duplexes considerable intra- and interstrand stacking exists. Not surprisingly then, in DNA:RNA hybrids we observed a similar distance dependence for the yields of intra- and interstrand CT [105]. So, the distinction between intra- and interstrand CT observed in B-DNA duplexes truly correlates with differences in base stacking.

3.2 Characterization of Transient Intermediates

The spectroscopic investigations described so far have largely been based on monitoring CT-induced quenching of photoexcited charge donors. These methods provide limited opportunity to detect the ions, radical ions or radicals that form as intermediates in charge transfer reactions. To enhance our view of DNA-mediated CT we have sought to directly observe and characterize transient intermediates along the CT pathway. In this endeavor we have exploited the flash quench scheme to generate the potent ground state Ru(III) oxidant in situ (Fig. 3). Reaction of the intercalated Ru(III) species with guanine or other bases in DNA has afforded us direct examination, by transient absorption spectroscopy, of the intermediates formed upon DNA-mediated CT. Furthermore, these studies have advanced our efforts to correlate CT intermediates with permanent chemical products.

The neutral guanine radical is detected immediately following nanosecond laser photolysis and flash quench generation of Δ-[Ru(phen)$_2$(dppz)]$^{3+}$ intercalated within poly(dG-dC) [59]. This radical is the product of electron transfer from guanine to the Ru(III) oxidant, followed by rapid deprotonation of the resultant guanine radical cation. Deprotonation is consistent with a pK_a of ~3.7 reported for the guanine radical cation in neutral aqueous solution [106]. These results are likewise consistent with the observation of the neutral guanine radical by EPR following flash quench experiments with poly(dG-dC) [107]. Formation of the guanine radical was found to be concomitant with quenching of photoexcited Δ-[Ru(phen)$_2$(dppz)]$^{2+}$ by the dif-

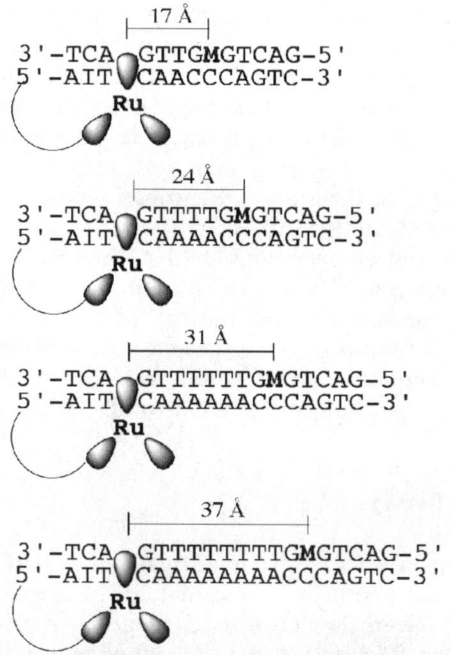

Fig. 8 Long range charge transport between dppz complexes of Ru(III) and an artificial base, methyl indole, in DNA. The methyl indole is paired opposite cytosine and separated from the intercalating oxidant by distances up to 37 Å. In all assemblies, the rate constant for methyl indole formation was found to be coincident with the diffusion-controlled generation of Ru(III) ($>10^7$ s^{-1}), indicating that charge transport is not rate limiting over this distance regime

fusible quencher (k_{obs}~2×10^7 s^{-1}). Therefore oxidation and deprotonation of guanine occur in less than ~200 ns in duplex DNA. The resultant guanine radical was observed to decay over hundreds of microseconds, resulting, ultimately, in the formation of permanent oxidation products as monitored by gel electrophoresis. Significantly, this investigation provided the first observation of the guanine radical in DNA by UV-vis spectroscopy, and established timescales for formation and deprotonation of the guanine radical cation as well as the decay of the guanine radical leading to permanent oxidative lesions.

In recent, analogous studies, we have directly observed a radical cation intermediate localized on a DNA base analog [42, 108]. Here the flash quench technique was employed to generate a Ru(III) oxidant intercalated within duplex DNA containing methyl indole as a hole trap. The resultant methyl indole radical cation was detected by both transient absorption and EPR spectroscopies. Corresponding irreversible oxidation of the indole moiety was demonstrated by biochemical assays that detected strand breaks at the indole sites following steady-state irradiation and treatment with piperidine. As for guanine, formation of the charge transfer intermediate was concomitant with the initial quenching of the photoexcited Ru(II) intercalator, $\geq 10^7$ s^{-1}. In fact, for the assemblies examined, containing no intervening guanines, CT to form the methyl indole radical cation was required over a distance of ~40 Å and was not rate-limiting (Fig. 8). Moreover, the yield of charge transfer intermediates and products is dramatically diminished by the presence of an intervening mismatch, and sensitively modulated by the extent of initial charge localization at the injection site [108].

These data provided the first direct measurements of rates of formation of transient radical species over long (>20 Å) molecular distances, and the first correlations between biochemical measurements of yield of oxidative damage and direct measurements of radical formation over this long range. Importantly, the facility in preparing these assemblies now will permit us to systematically vary sequence and distance in order to directly test various mechanistic proposals for long-range CT through DNA.

3.3 Long-Range Oxidative Damage

In the course of our spectroscopic investigations of DNA-mediated CT, we were naturally curious about the consequences of electron and hole migration through DNA. Given that radicals damage DNA within the cell, could these fast, long range CT reactions be physiologically relevant? Could DNA-mediated CT induce chemistry, in particular oxidative damage to DNA, from a distance?

Our first biochemical experiments addressing these questions revealed that indeed CT through DNA induces chemistry at distance [109, 110]. These investigations probed the migration of positive charges, or holes, though DNA. Since G has the lowest oxidation potential of the four DNA bases [39], migrating holes are expected to be trapped at G, leading, potentially, to permanent oxidative damage. Furthermore, the 5′-G of 5′-GG-3′ doublets has been found to be particularly sensitive to oxidation [111]. This has been rationalized by ab initio molecular orbital calculations which predict that the oxidation potential of 5′-GG-3′ doublets is reduced (relative to single guanines) and that the HOMO lies predominantly on the 5′-G [90]. Preferential oxidative damage at the 5′-G of 5′-GG-3′ doublets has therefore become a signature of CT chemistry.

To examine the chemistry of migrating holes in DNA, we prepared DNA assemblies containing a tethered intercalating photooxidant, $[Rh(phi)_2 DMB]^{3+}$, spatially separated from two GG sites [109]. Upon irradiation with 365 nm light, we observed oxidative damage to both GG sites, 17 and 34 Å away from the photooxidant. Photolysis at 313 nm confirmed that the rhodium complex intercalated near the terminus of the duplex where it was tethered, and that no inter-assembly intercalation had occurred. Remarkably, the yield of oxidative damage was essentially the same at both GG sites. The damage yields were, however, modulated by changes in oxidation potential at the guanine sites, or changes in stacking of the intercalator within the DNA helix. Stacking of the intervening DNA bases was also found to be crit-

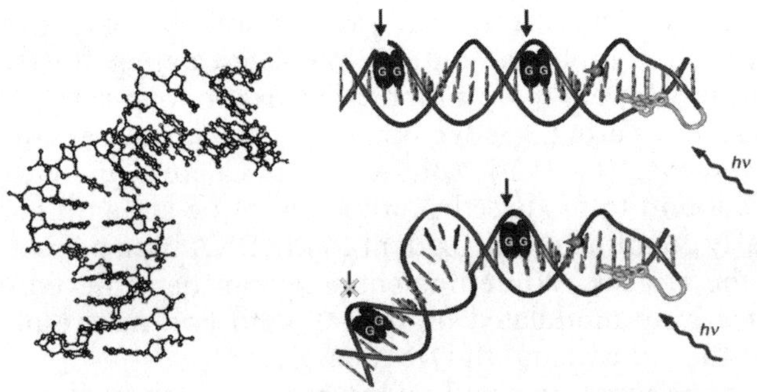

Fig. 9 Long-range oxidative damage in DNA is regulated by the structure of the DNA π-stack. Photoexcited $[Rh(phi)_2(bpy')]^{3+}$ intercalated within B-DNA oxidizes guanine doublets positioned 17 Å and 40 Å away with approximately equal efficiency. When a bulge is introduced in the DNA oxidation at the distal guanine doublet is inhibited. As shown by the NMR structure, bulges introduce local distortions in the DNA π-stack, although the kinking of the helical actually positions the distal guanine closer to the Rh(III) photooxidant

ical; DNA assemblies possessing base bulges between the 5′-GG-3′ sites distal and proximal to the intercalated photooxidant displayed a dramatic reduction in the distal/proximal ratio of oxidative damage [110] (Fig. 9). Undoubtedly this oxidative chemistry occurs by hole migration *through* the DNA base stack. These seminal experiments therefore established that long range HT leads to permanent oxidative damage at guanine, and that this chemistry is characterized by a pronounced sensitivity to stacking with minimal sensitivity to distance.

To systematically examine the distance dependence of long range oxidative damage, we constructed a series of 28 base pair duplexes containing guanine doublets distal and proximal to a tethered Rh(III) intercalator [112]. With the position of the proximal GG doublet fixed, the separation between the photooxidant and the distal GG site was increased in 2 base pair increments. Consistent with our spectroscopic measurements of CT, the yield of oxidative damage was not significantly attenuated over a distance of 75 Å. In addition, the helical phasing of the intercalator and the site of oxidation (for instance stacking of the intercalator and GG doublet on the same or opposite sides of the helix) did not appear to significantly influence the CT yield.

Oxidative damage to guanine via long range DNA-mediated hole transport is not specific to our Rh(III) intercalators. We have observed this chemistry using several other intercalating oxidants, include ground state Ru(III) species [112, 113], and ethidium [114]. Indeed, the generality of long-range guanine oxidation via DNA-mediated HT has now been established by investigations of several different laboratories using a variety of distinct oxidants [115–119]. Importantly, we have consistently observed that the ability of the oxidant to initiate damage at a distance is fundamentally related to its association with the DNA base stack. This essential feature of DNA-mediated HT was unambiguously demonstrated in a recent study of oxidative damage by a family of Ru(II) complexes [120]. While intercalating Ru(II) complexes, such as $[Ru(bpy)_2(dppz)]^{2+}$, promote significant long-range oxidation of guanine, non-intercalating, groove binding species, such as $Ru(bpy)_3^{2+}$, are ineffective. Moreover, for a series of Ru(II) intercalators, the yield of guanine oxidation was found to be directly correlated to the strength of intercalative binding. Clearly coupling of the oxidant to the DNA base pairs is requisite to HT through the π stack. While this coupling can be achieved by intercalation, it is sensitively modulated by the strength and nature of the interactions between the oxidant and the DNA bases.

Having demonstrated, generally, long-range HT through the DNA base stack with a shallow dependence on distance, an obvious question arises: how far can holes migrate in DNA? In particular, can HT generate oxidative damage over physiologically relevant distances? We tested this notion by probing damage to GG doublets positioned up to 200 Å from either a photoexcited Rh(III) or a ground state Ru(III) intercalating oxidant [112]. Remarkably, HT reactions yielding significant oxidative damage were observed over these biologically important distances. Such long-range oxidative damage to guanine has also been observed using anthraquinones as photooxidants [121]. While long-range HT did exhibit some dependence on distance, we found the intervening sequence and sequence dynamics to play a significant role in the efficiency and distance dependence of oxidative damage. For instance, the yield of oxidative damage was diminished through multiple 5′-TA-3′ steps, and an increase in the proportion of long-range damage observed as the temperature was increased from 5 to 35 °C. These observations reveal important features of very long distance DNA CT. These features coincide with our picture of shorter range CT developed from spectroscopic studies, and may be significant to DNA damage in vivo.

Once long range oxidative damage in DNA had been demonstrated, our attention shifted towards understanding *how* holes migrate through DNA over such distances. Consistently, our investigations of CT reactions indicated that stacking interactions involving the DNA bases and the redox participants would be at the center of any mechanistic description of CT in DNA. We have performed a variety of experiments to test this notion and to enhance our understanding of the mechanisms of long-range guanine oxidation.

Although we tend to visualize DNA in its standard B-form, DNA exhibits incredible structural diversity [7, 122, 123]. Long range guanine oxidation has now been shown to occur in an array of DNA structural forms including single-strands [124], triplexes [125, 126], cross-over junctions [127, 128], DNA:RNA hybrids [129, 130] and Z-DNA [131]. A wide range of structures and structural dynamics can be accessed by exploiting DNA polymorphism. We have investigated non B-form DNA to probe how the structure of bases within DNA helices is related to charge transport function (Fig. 10).

In DNA:RNA hybrids, where intrastrand base stacking is comparable to B-DNA, we observed efficient long range guanine oxidation when using ethidium as an intercalating photooxidant [130]. Conversely, a Rh(III) com-

Fig. 10 Charge transport is observed in a variety of nucleic acid assemblies over a wide distance regime (3.4–200 Å). Shown are examples of nucleic acid structures through which charge transport has been examined: **a** B-form DNA; **b** DNA-RNA hybrids; **c** cross-over junctions; and **d** nucleosome core particles. In all assemblies, the charge transport chemistry is extremely sensitive to the structure of the π-stacked nucleic acid bases

plex, that promotes guanine oxidation at a distance in B-DNA, was found to be ineffective in DNA:RNA hybrids. Here, the weakened intercalative binding of the Rh(III) complex within the A-form DNA:RNA hybrid proved critical; poor coupling of the photooxidant with the DNA base stack reduces the efficiency of charge injection and consequently long range chemistry. Similarly, efficient long range oxidative damage in DNA triple helices was observed only when the photooxidant was intercalated within the center of the triplex [126]. Due to the distorted stacking at the 5′ end of the triplex within the duplex-triplex junction, a photooxidant tethered to the 5′ end of the triplex failed to oxidize distant guanines. Again, coupling of the intercalator to the DNA base stack was found to be essential to long range HT.

Long range guanine oxidation was also observed as a result of HT through a single cross-over junction assembled from four partially complementary DNA strands [127]. Radical migration throughout the assembly was attributed to the flexibility of the single cross over junction. Rapid sampling of a variety of conformations thereby facilitates transient base stacking in normally disfavored arrangements. This notion was supported by experiments in double crossover assemblies. In these assemblies, which are even more rigid than B-DNA, radical migration occurred selectively down the base stack bearing the intercalated photooxidant [128].

Non B-form DNA assemblies undoubtedly induce rather dramatic modifications in the structure and dynamics of the DNA bases. More subtle differences in base stacking can be achieved in B-DNA as a function of base sequence. We have repeatedly noted distinctions in the rates, yields and distance dependence of CT reactions as a function of DNA sequence. In a systematic investigation of the sequence dependence of long-range guanine oxidation, we employed 21 base-pair DNA duplexes possessing guanine doublets distal and proximal to a tethered $[Rh(phi)_2bpy']^{3+}$ photooxidant [132].

The sequence immediately surrounding the guanine doublets was fixed, while the length and arrangement of the intervening A/T bridge was modulated. Consistent with expectations derived simply from base stacking, we observed the most efficient long-range oxidation through adenine bridges. The ability of attenuating A/T bridges to mediate HT was substantially lower, while thymine bridges displayed an intermediate efficacy. Perhaps more enlightening, however, was the fact that the yield of damage to the distal guanine doublet actually *increased* with the length of the intervening A/T bridge. Furthermore, incorporation of a G/C step within the A/T bridge decreased the efficiency of long-range HT.

These results underscore the complex role of sequence dependent structure and dynamics in DNA-mediated CT. Certainly the observations cannot be rationalized by models in which holes hop along guanines via superexchange through short A/T bridges [133, 134]. Nor are the results consistent with a "zig-zag" mechanism involving indiscriminate intra- and interstrand migration that is not influenced by A-T base-pair orientation [133, 135]. Here, the increase in HT efficiency with increasing bridge length was attributed to the formation of transient, well-coupled conformations, distinct from canonical B-DNA. Such structures are known to arise in A-tracts, once nucleated by a sufficient number of adenines [136]. The inclusion of the G/C step within the otherwise A/T bridge disrupts the formation of these transient conformations. Interestingly, models based on thermally-induced hopping also predict that inclusion of a G/C step in a long A/T bridge should decrease the efficiency of HT [137]. Likewise, the particular flexibility of 5′-TATA-3′ steps [138] could certainly be related to the reduced efficiency of HT through alternating A/T bridges.

Modulation of DNA structure and dynamics is also possible using base-pair mismatches. Mismatches exert little influence on the global structure of B-DNA duplexes. Locally, the extent of base stacking perturbation depends sensitively on the nature of the mismatch [139–141]. Therefore, while a CA mismatch introduces a significant distortion in local stacking, the well-stacked GA mismatch is, by many criteria, barely perceptible. The dynamics of mismatched base-pairs may also be significantly distinct from matched Watson-Crick base pairs [9]. We exploit these features of DNA mismatches to probe the sensitivity of DNA-mediated CT to base structure and dynamics.

Both our spectroscopic investigations [64] and our biochemical assays [113] revealed a diminution in CT efficiency in the presence of mismatches which disrupt the DNA π-stack. Recently we conducted a systematic investigation of long range HT through DNA as a function of intervening base mismatches [142]. In these experiments we probed oxidative damage in 22 base-pair DNA duplexes containing mismatches between guanine doublets distal and proximal to a $[Ru(phen)(dppz)(bpy')]^{2+}$ intercalator. The extent of distal guanine oxidation was compared with the helical stability, electrochemical measurements of intercalator reduction on different mismatch-containing DNA films, and the lifetimes of the mismatched base pairs determined from ^1H NMR measurements of imino proton exchange rates. Significantly, the extent of long-range oxidative damage correlated most closely with the base-pair lifetimes of the intervening mismatches. Furthermore, competitive hole trapping at the mismatch site [143] did not modulate the efficiency of HT through the mismatch. Once again base-pair dynamics are shown to govern the efficiency of long range HT.

Investigations of guanine oxidation at a distance strongly suggest that hole migration through well-stacked DNA is significantly faster than trapping of the resultant guanine radical by oxygen and/or water. In particular, our distal/proximal ratios of oxidative damage support a mechanism in which equilibration of the hole is rapid on the timescale of trapping. With such a mechanism, similar yields of oxidation are expected for all guanine doublets on a given DNA duplex, provided the thermodynamic potentials are approximately equal. Interestingly, our Rh(III) and Ru(II) intercalators often yield distal/proximal damage ratios that are significantly *greater* than one. Could the high cationic charge of the metallointercalator bound near one end of the duplex attenuate the potential of the proximal guanine doublet? Indeed, investigations of the influence of ionic distribution on HT suggest that this may be the case [144]. Here oxidative damage was monitored in DNA assemblies constructed of a DNA strand functionalized at the 5′-end with $[Rh(phi)_2bpy']^{3+}$, and a complementary strand radiolabeled with negatively-charged ^{32}P at either the 5′ or 3′ end. Moving the negatively-charged label from the 5′ to the 3′-end substantially diminished the yield of oxidative damage to the distal site. Likewise, the inclusion of negatively-charged phosphates at both the 5′ and 3′ ends of the complementary strand resulted in an intermediate distal/proximal damage ratio. We rationalized these results in terms of changes in the oxidation potential at the distal and proximal guanine sites induced by changes in charge at the termini of the duplex. Certainly these intriguing results illustrate the complex interplay of parameters that influence charge migration through DNA in solution and must be considered in the development of mechanistic schemes.

The fact that the ionic environment of DNA should influence migrating charges is not particularly surprising. Nonetheless, it was suggested that our observations concerning the influence of ionic distribution were an artifact of our Rh(III) intercalators, and were unrelated to mechanisms of CT through DNA [145]. This suggestion was based on the fact that an analogous influence of terminal charge distribution was not detected using an end-capped anthraquinone as a photooxidant. However, based on the very low distal/proximal ratios observed with the anthraquinone photooxidant, it is immediately evident that such an effect *should not* be seen in these assemblies. The influence of ionic distribution we observed relies on the fact that charge equilibration is rapid on the time-scale of charge trapping. Consequently the yield of oxidative damage is governed by the thermodynamic potentials. Rapid charge equilibration is revealed by distal/proximal ratios which are unity, or greater. With the anthraquinone photooxidant, distal/proximal ratios are notably less than one, indicating that the charge may not be equilibrated and the yield ratio should not be determined by the potentials at the guanine sites. Clearly, then, Rh(III) intercalators are unique in that they are very strongly coupled to the π-stack thereby facilitating effective charge injection. Consequently, Rh(III) intercalators accurately monitor mechanisms of CT through the DNA helix.

3.4
Oxidative Repair of Thymine Dimers at a Distance

While long-range oxidative damage to DNA has now been extensively investigated, the potential of DNA-mediated CT to induce other chemical reactions at a distance remains largely unexplored. A particularly attractive possibility is that CT through DNA may also induce DNA repair. Thymine dimers are the most prevalent photochemical lesions in DNA. These cyclobutane dimers form as a result of a photoinduced [2+2] cycloaddition between adjacent thymines on the same polynucleotide strand. While eukaryotic cells excise the thymine dimer, in bacteria the lesion is repaired by photolyase via electron transfer from a reduced flavin cofactor to the cyclobutane dimer [146]. Model studies have demonstrated that the thymine dimer can also be repaired oxidatively (E^0 T<>T~2 V vs NHE) [147, 148]. Consequently, thymine dimer repair is a physiologically significant reaction which can be accomplished by electron or hole transfer with no additional chemistry. Our Rh(III) complexes, as powerful photooxidants ($E^*(Rh^{3+/2+})$~2 V vs NHE) and avid intercalators, represented excellent candidates for oxidative repair of thymine dimers at a distance.

We first demonstrated long-range oxidative repair of thymine dimers in 16 base-pair DNA duplexes functionalized with an intercalating $[Rh(phi)_2 bpy']^{3+}$ photooxidant 16 to 26 Å away from the dimer lesion [149] (Fig. 11). Using high performance liquid chromatography (HPLC) we monitored the progressive repair of the dimer with increasing irradiation time at 400 nm. As for long-range oxidative damage, this chemistry was largely insensitive to distance, but readily perturbed by disruptions of the intervening π-stack. Therefore the same DNA-mediated HT which induces long-range oxidative damage could also promote oxidative repair at a distance.

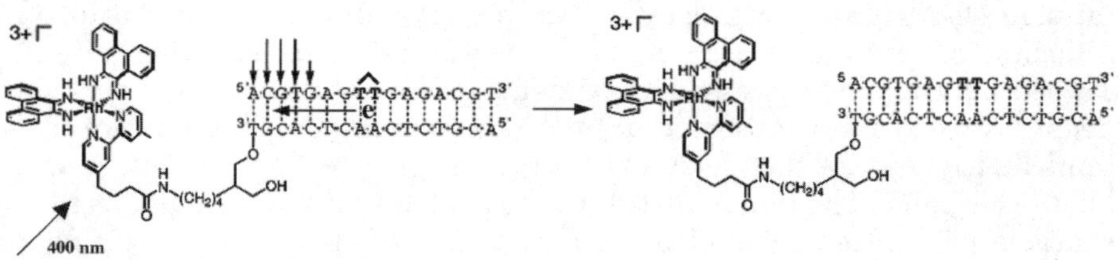

Fig. 11 Thymine dimer repair at a distance by DNA-mediated charge transport. Here photoexcitation of intercalated $[Rh(phi)_2bpy']^{3+}$ tethered to the 3′-end of a DNA duplex oxidizes a remote thymine dimer (≥34 Å away) within the helix leading to dimer repair. The arrows mark the sites of intercalation of the phi ligand. Adapted from [149]

In light of these observations, we were interested in examining the competition between long range oxidative damage and repair, particularly given the fact that guanine oxidation should be thermodynamically favored by ~0.7 V relative to thymine dimer repair. We evaluated the reactivities of thymine dimers and guanine doublets within the same duplex covalently tethered with intercalating oxidants, either $[Rh(phi)_2bpy']^{3+}$ or $[Ru(phen)(dppz)bpy']^{2+}$ [150]. The Ru(III) oxidant did not repair the thymine dimer, consistent with the fact that the dimer is not thermodynamically accessible. It did, however, oxidize guanine doublets on *both sides* of the dimer. In Rh(III) modified duplexes, both guanine oxidation and thymine dimer repair are observed; thymine dimer repair proceeds efficiently, despite the presence of the guanine doublets as thermodynamic traps. These results indicate that it is kinetically favorable to repair the thymine dimer. The competition between GG oxidation and thymine dimer repair in vitro thus appears to be associated with relative kinetic versus thermodynamic control of these reactions. Understanding the competition between these CT reactions in vivo will certainly shed some light on the consequences and regulation of DNA CT chemistry in biological systems.

Oxidative repair is not a unique feature of our Rh(III) complexes. We also demonstrated efficient long-range repair using a covalently tethered naphthalene diimide intercalator ($E^{*-/0}$~1.9 V vs NHE) [151]. An intercalated ethidium derivative was ineffective at dimer repair, consistent with the fact that the reduction potential of Et^* is significantly below the potential of the dimer. Thymine dimer repair by a series of anthraquinone derivatives was also evaluated [151]. Despite the fact that the excited triplets are of sufficient potential to oxidize the thymine dimer ($E^{3*-/0}$~1.9 V vs NHE), the anthraquinone derivatives were unable to effect repair [152]. We attribute the lack of repair by these anthraquinone derivatives to their particularly short-lived singlet states; anthraquinone derivatives that do not rapidly interconvert to the excited triplet state are indeed effective at thymine dimer repair [151]. These observations suggest that interaction of the dimer with the singlet state may be essential for repair.

3.5 Charge Transport Through DNA Films

Through our spectroscopic investigations and biochemical assays we have extensively characterized DNA-mediated CT, including long-range oxidative damage and repair. In all cases, photoexcited probes initiate CT through the base stack of DNA molecules in aqueous solution. In order to broaden our perspective of the mechanism and scope of DNA CT, we exploited the ability of DNA for self assembly on gold surfaces via alkane-thiol linkers. The resultant DNA-modified surfaces represent novel media in which to investigate DNA-mediated CT reactions; CT reactions involving ground state probes can be interrogated electrochemically, thereby facilitating characterization of DNA CT in still more energetic and time regimes.

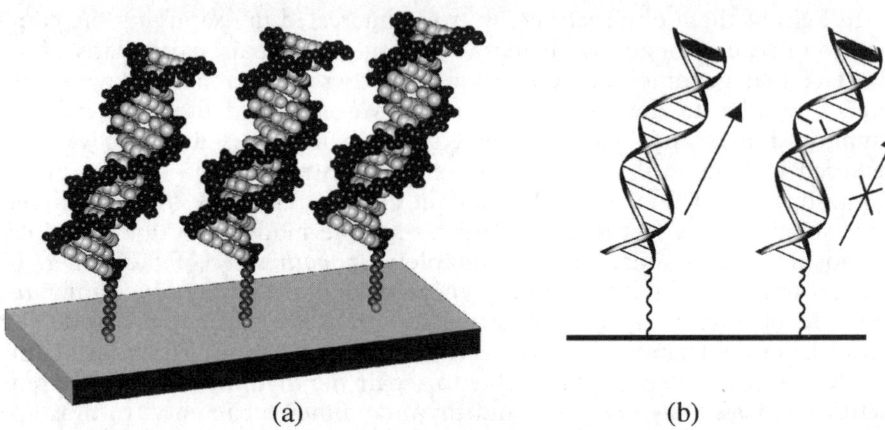

Fig. 12 DNA-mediated charge transport through DNA films. **a** Schematic representation of alkane thiol-modified DNA duplexes self-assembled on gold electrodes. **b** As in solution, charge transport through DNA films is strongly dependent on minor perturbations in the DNA π-stack. For instance, a single base mismatch switches off the current flow from the gold electrode to a DNA intercalator bound near the top of the duplex. Consequently charge transport through DNA is a sensitive reporter of DNA structure, including single-base mismatches, and structural changes induced upon protein binding

Our DNA-modified surfaces are typically constructed of small (e.g. 15 base pair), tightly packed DNA duplexes on gold electrodes (Fig. 12). Immobilization of a high density of DNA is achieved through self assembly of alkane-thiol functionalized duplexes in the presence of a high concentration of magnesium ion. Our picture of the DNA-modified surfaces is derived from extensive spectroscopic and biochemical characterization [92, 153, 154]. For instance, the surface density has been directly evaluated using ^{32}P radioactive labeling, while the morphology of the films has been uncovered using atomic force microscopy (AFM). These AFM studies revealed surfaces densely covered with a monolayer 45 Å in thickness, indicating that the DNA duplexes are oriented at an angle of ~45° relative to the gold surface. Furthermore, in AFM studies as a function of potential, we observed that application of a positive potential causes the DNA, if not tightly packed, to lie down on the surface, thereby compressing the monolayer (20 Å). Conversely, application of a negative potential increases the monolayer thickness (50 Å), consistent with the DNA duplexes oriented perpendicular to the surface.

These observations were significant to our choice of reactants for probing CT at DNA-modified surfaces. In particular, an upright orientation of the DNA relative to the surface is required to probe DNA-mediated reactions; otherwise a more direct reaction between an intercalating probe and the electrode might be possible. Consequently, reactants were selected such that a negative potential could be applied, thereby initiating reduction of an intercalated redox probe distantly bound within DNA helix. Importantly, the resulting DNA-mediated process therefore involves electron rather than hole transport.

In our first investigations we observed reduction of micromolar concentrations of the organic intercalator, methylene blue (MB), on an electrode modified with densely packed DNA duplexes [92]. While the binding affinity of MB to surface-bound DNA was found to be comparable to DNA in solution, the binding stoichiometry was significantly reduced, consistent with MB accessing only sites near the top of the densely-packed DNA film. If so, the distance between the gold surface and the MB was significant, and these first experiments suggested long-range DNA-mediated electron transport.

To unambiguously test the influence of distance on this DNA-mediated electron transport, we prepared a series of thiol-modified DNA duplexes containing a single G-C step functionalized with DM [93]. DNA films constructed of DM-modified duplexes were shown to be structurally indistinct from DNA films prepared with duplexes lacking the crosslinked moiety. Rates of electron transport through the DNA films were estimated in cyclic voltammetry experiments where the scan rate was varied. Remarkably, although the position of the intercalating DM was varied by 40 Å, no change in the rate of reduction was detected; slow rates of 10^2 s^{-1} were consistently observed. Comparable rates have been reported for reduction of other redox-active probes directly attached to alkane-thiol linkers [94]. These observations suggest that the rate-limiting step for reduction of DM bound to DNA-modified surfaces is tunneling of the electron through the alkane-thiol linker. Consequently, electron transport through the DNA helix over a distance of at least 40 Å is necessarily faster. Equally important was our observation that the reduction of DM was effectively "turned-off" by the incorporation of a C-A mismatch within the DNA duplex (Fig. 12). This confirmed that the pathway for electron transport was *through* the DNA base stack. Furthermore, it illustrated that π-stack perturbations within DNA films sensitively modulate electron transport, consistent with our picture of CT through DNA in solution. Interestingly, electron transport through DNA films does not display the dramatic dependence on sequence observed for DNA-mediated CT in solution [155], perhaps because the sequence-dependent flexibilities of DNA are tempered by positioning within the close-packed DNA film.

4 Applications and Biological Consequences of DNA-Mediated Charge Transport

While CT mediated by the DNA base stack may not be completely characterized mechanistically, that it does occur is undeniable. This remarkable chemistry, inherent to DNA, has opened the door to a myriad of possible applications, and provoked intriguing questions regarding biological consequences.

4.1
DNA Charge Transport as a Probe of π-Stacking Perturbations

The innate sensitivity of DNA-mediated CT to perturbations in the π-stack has prompted us to employ this chemistry as a probe of stacking structure and dynamics. We have developed a new class of DNA-based diagnostic tools that diagnose DNA mutations such as single base-pair mismatches and lesions, analyze DNA-protein interactions, and probe the sequence-dependent dynamics and flexibility of DNA. These applications rely on electrochemical probing of CT in DNA films self-assembled on gold electrodes.

Our initial investigations of CT at DNA-modified surfaces demonstrated that a C-A mismatch effectively short-circuits the current through the DNA duplexes. This confirmed that the pathway of CT was through the helix. However, it also revealed that a single-base mismatch could be accurately detected electrochemically. Consequently we have exploited this CT chemistry to fabricate a device for sensitive mismatch detection [155, 156]. By amplifying the sensitivity with an electrocatalytic cycle, all possible mismatches, including purine-purine pairings (such as GA), which induce only small stacking perturbations, can be detected, as can a number of naturally occurring lesions (for instance A_{ox}:T). Mutations and lesions have also been detected in physiologically relevant sequences, such as in the p53 gene, where mutations implicated in a variety of human cancers are known to occur. Furthermore, this CT chemistry is compatible with chip-based technology; we have developed DNA chips using gold surfaces of variable size on silicon wafers that facilitate the detection of as few as 10^8 DNA molecules per electrode.

Charge transfer through DNA films has proven to be a sensitive probe of many other types of π-stacking perturbations. For instance, with this technology we have examined the influence of nucleic acid analogs, such as locked nucleic acids used in antisense applications, on duplex stacking [157]. Various levels of π-stack perturbations induced by DNA-protein interactions are also readily and sensitively assessed by electrochemically probing CT through DNA films [158]. Here thiol-terminated DNA duplexes functionalized with a crosslinked DM are self-assembled on gold electrodes in the absence of magnesium ion. This ensures loose packing of the DNA film thereby allowing protein binding; back-filling with mercaptohexanol passivates any gold surface which remains exposed. Electrochemical interrogation of DNA-protein interactions consistently reports on the degree of π-stack perturbation induced by protein binding. For instance, binding of M.*Hha*I, which flips a DNA base out of the π-stack and inserts a glutamine residue in its place, resulted in a dramatic reduction in the electrochemical response of the DNA film. Significantly, binding of a mutant enzyme (Q237W), which instead inserts the aromatic tryptophan moiety into the π-stack, caused little reduction of the electrochemical signal. Binding of uracil-DNA glycosylase (UDG), another base flipping enzyme, or TATA-binding protein (TBP), which induces a dramatic structural distortion, similarly induced a sharp attenuation of the electrochemical response.

These studies nicely complement our investigations of DNA-protein interactions based on measurements of long-range guanine oxidation in solution (vide infra), despite the fact that here we were measuring, electrochemically, the reduction of a bound intercalator. Furthermore, electrochemical measurements of CT through DNA films represent a sensitive and facile experimental technique for monitoring DNA-protein interactions and enzymatic reactions in real time. For example, investigation of the endonuclease activity of *R.Pvu*II on a DNA films observed a decrease in electrochemical response as a function of incubation time with the protein; the corresponding kinetics closely matched those determined in a gel assay of the restriction fragments [158]. Remarkably, these valuable diagnostic tools originate from the incredible sensitivity of DNA-mediated CT to static changes and dynamic fluctuations in base stacking.

4.2
DNA Charge Transport In Vivo

Despite the wealth of experimental and theoretical investigations of CT in DNA assemblies designed by chemists and physicists, the role of DNA CT in biology remains to be discovered. Does long-range DNA CT occur in vivo? If so, why, and how is it regulated? How do proteins influence and/or participate in DNA-mediated CT? We have initiated a variety of experiments to address these fundamental questions.

An important consideration for DNA CT in the cell is the dramatically different environment of the DNA molecule in vivo. Unlike the naked DNA typically used in our in vitro assemblies, cellular DNA is intimately associated with proteins, packaged into chromosomes, and stored in the cell nucleus. Several experimental investigations demonstrate definitively that proteins can both modulate and participate in CT reaction in DNA.

As observed in electrochemical investigations, long-range guanine oxidation via DNA-mediated CT is disrupted by proteins which perturb the structure of the DNA base-pairs [159, 160]. We constructed DNA assemblies that contained the protein binding site between two guanine doublets spatially separated from a tethered $[Rh(phi)_2bpy']^{3+}$ photooxidant. DNA-binding proteins such as M.*Hha*I or TBP, which disrupt DNA base stacking by base flipping or dramatically kinking the DNA, respectively, decrease the yield of guanine oxidation upon binding. A remarkable testament to the importance of base stacking is the fact that long-range CT chemistry is restored by the mutant M.*Hha*I (Q237W) protein which inserts an aromatic tryptophan residue within the π-stack. Alternatively proteins such as the restriction endonuclease, *R.Pvu*II, and the transcription factor, Antennapedia homeodomain protein (ANTP), which do not disrupt the π-stack, but may rigidify the helix upon binding, enhance the ability of DNA to mediate long-range CT.

These results, completely analogous to our electrochemical assays, illustrate how protein binding sensitively modulates long-range DNA-mediated CT in a manner that is highly dependent on the specific DNA-protein interactions involved. Consequently, we have recently applied this chemistry as a probe of DNA-protein interactions. For instance, we have examined the MutY/DNA interaction [161]; MutY is a DNA repair enzyme found in *Escherichia coli*. Here no disruption of DNA-mediated CT is detected upon protein binding, suggesting that MutY does not locate its binding site by progressive base flipping.

Having established that proteins influence DNA-mediated CT, we were curious as to how proteins might *participate* in CT reactions involving DNA. In particular, can electrons and holes be transferred between DNA and proteins, and can these reactions be triggered at a distance? Spectroscopic and biochemical assays have therefore been employed to probe CT reactions in protein/DNA assemblies. In first experiments, using the flash quench technique, we observed long distance oxidation of both tyrosine and tryptophan when Lys-X-Lys (X=tyrosine or tryptophan) tripeptides were bound to DNA [162, 163]. These results established direct participation of the DNA-bound peptides in long-range CT reactions through DNA, and prompted questions regarding the functional role of proteins in these CT reactions.

Subsequently we examined both the transient intermediates and the yields of oxidative damage induced upon flash quench generation of an Ru(III) oxidant intercalated within DNA containing the M.*Hha*I binding site [164]. Upon irradiation of an assembly with the mutant M.*Hha*I (Q237W) bound 14 base-pairs away from the Ru(III) oxidant, significant oxidative damage was detected at the guanine located 5′ to the site of tryptophan intercalation. Importantly, no such damage was evident upon binding of the wild type enzyme which inserts a glutamine into the DNA base stack. In transient absorption experiments, both the guanine and tryptophan ($E^{0/+}$~1 V vs NHE) radicals were detected as intermediates. Therefore, we have directly observed long-range hole transport through DNA to the inserted tryptophan of a bound protein and to the adjacent 5′-guanine yielding permanent oxidative damage (Fig. 13).

Interestingly, the yield of oxidative damage decreased only slightly with increasing distance (24–51 Å) between the M.*Hha* I binding site and the intercalated Ru(III) oxidant, while the rate constant for radical formation was unchanged (>10^6 s^{-1}), irrespective of distance. Consequently, HT through DNA over this 50 Å distance regime is not rate-limiting. This conclusion is supported by recent investigations employing the artificial base, methyl indole, directly incorporated within the DNA base stack [42, 108]. Comparable results were observed in these studies, and the lower limit for CT through the DNA base stack was set at 10^7 s^{-1}, coincident with diffusional quenching to generate our Ru(III) oxidant.

Other intriguing questions relate to the possibility of electron transfer from a protein to oxidatively damaged DNA. We have probed this possibility by examining electron transfer from ferrocytochrome C to a guanine radical in DNA [165]. The flash quench technique was used to generate the potent ground state oxidant, $[Ru(phen)_2dppz]^{3+}$, intercalated within poly(dG-dC). Transient absorption measurements revealed rapid oxidation of guanine by the Ru(III) intercalator, followed by slower electron transfer from the ferrocytochrome C to the resultant guanine radical. Equally intriguing has been

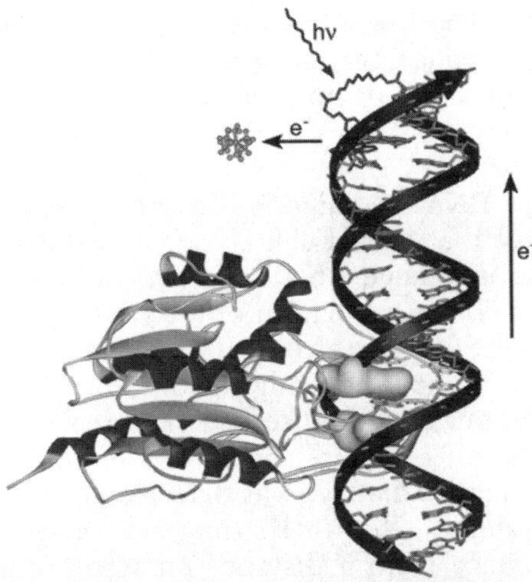

Fig. 13 DNA-protein CT reactions. The DNA-bound protein, methyltransferase *Hha*I (mutant Q237W), flips a base out of the DNA double helix and inserts a trytophan side chain leaving the π-stack largely unperturbed. This intercalated trytophan moiety transfers an electron to $[Ru(bpy')(dppz)(phen)]^{3+}$, generated by flash quench, over 50 Å away. Adapted from [164]

our recent observation of DNA-mediated CT to the [4Fe-4S] cluster of the DNA repair protein MutY [166]. It has been demonstrated that binding of this repair protein to DNA activates the cluster towards oxidation. Therefore DNA repair via DNA-protein electron transfer has been revealed as a fascinating area for future investigations.

As it became evident that proteins both modulate and participate in CT reactions through DNA, questions arose concerning DNA CT in vivo, where the DNA is intricately associated with proteins. In eukaryotic cells, DNA is packaged in nucleosome core particles; ~150 base pairs of DNA are coiled around an octamer of histone proteins to which the DNA is associated by non-specific electrostatic interactions. Would CT proceed through DNA within the nucleosome core particle? Although the DNA is less accessible to damaging reagents within solution, oxidative damage through DNA-mediated CT does not require such accessibility. However, the restricted motion and overall bending of the DNA within the nucleosome might disrupt long-range CT.

We investigated DNA-mediated CT through nucleosome core particles (Fig. 10) using photoexcited Rh(III) intercalators, either non-covalently bound or covalently tethered to the 146 base-pair DNA duplexes packaged into nucleosomes [167]. Although the histone proteins inhibited intercalation of the rhodium complex within the core particle, they did not prevent oxidative CT through the DNA over distances up to 75 Å. Histone binding did not influence the extent or pattern of oxidative damage. However, guanine oxidation was not detected at more distant sites even in the absence of the histone proteins; it is likely the bent DNA structure used to generate consistent nucleosome phasing interfered with CT. Significantly, then, although packing of the DNA in nucleosomes may provide protection from solution-mediated damage, it does not protect against oxidative damage via DNA-mediated CT.

To further explore DNA-mediated CT in the physiological realm, we explored the possibility that such CT may occur within cell nuclei [168]. In first experiments we irradiated Rh(III) intercalators which had been incubated with HeLa nuclei. After isolation of the DNA and treatment with base excision repair enzymes to reveal oxidative damage as strand breaks, the damaged DNA was amplified the using ligation-mediated polymerase chain reaction. Oxidative damage was observed preferentially at the 5′-G of 5′-GGG-3′, 5′-GG-3′, and 5′-GA-3′ sites, indicative of DNA-mediated CT chemistry. Remarkably, oxidative damage was detected at sites that, due to protein binding, are inaccessible to the Rh(III) intercalator. So, DNA-mediated CT can induce base damage from a distance on transcriptionally-active DNA within the cell nucleus. These exciting observations have propelled studies of in vivo DNA CT to the forefront, and certainly must be considered as we delve into cellular mechanisms for DNA damage and repair.

5
A Mechanistic View: Commonalities and Outstanding Issues

The donor-DNA assemblies presented in Table 2 have been used to probe CT over distinct regimes of time, energetics and coupling. Is it possible to reconcile these many different experiments together within one mechanistic picture? Many different theoretical treatments have been proposed, and to greater or lesser extents, these can be used to understand aspects of CT within DNA. However, many outstanding issues remain. Some are highlighted here.

We consider charge transport in various distance regimes. Over the short range mechanistic descriptions have focused primarily on superexchange with charge transfer occurring without occupation of the DNA bridge. In these cases, however, it remains critical to establish donor acceptor energetics relative to the bridge and the donor bridge coupling. Only when the energy gap is much larger than the coupling, can a superexchange description be strictly applied. In most cases these parameters are not known with certainty. At the other end of the mechanistic continuum is multistep hopping with true occupation of bridge states. Single step CT would require effective coupling throughout the base pair stack, over long molecular distances. The probability of having no transient structural defect in the stack is vanishingly small and hence very long range CT likely entails a multistep process, requiring occupation of the bridge states. Here too, however, many parameters may vary that are not known with certainty. These include the sequence-dependent structure and dynamics of the DNA bridge, the rate of charge injection into the bridge, and the extent to which charge is delocalized within the bridge. Moreover, what is the distance range over which real versus virtual

Table 2 Oxidants used to probe DNA-mediated charge transport, and features of charge transport with each oxidant

Oxidant[a]	Ref.[b]	Structural Details	Potential (V vs NHE)	τ
Rh(phi)$_2$(bpy')$^{3+*}$ 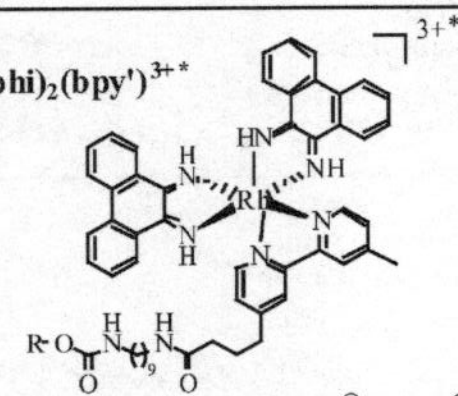	44, 50, 51, 112, 132, 144	covalently tethered to 5'-terminus of DNA with flexible 9-carbon linker; intercalates between DNA bases; no global or local base stack distruptions 1.2 Å crystal structure of Rh(III) phi complex intercalated into DNA; high resolution NMR	E*(Rh(III)/(II)) 2.0	100-400 ns
Ru(dppz)(x)(y)$^{3+}$ 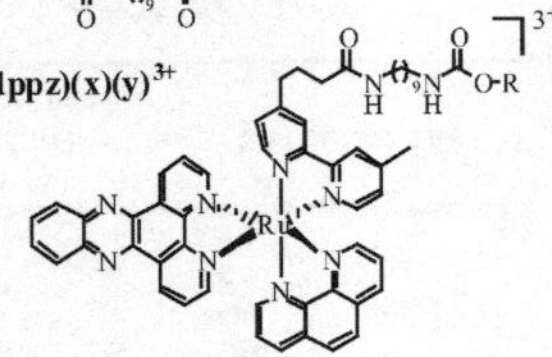	42, 44, 55, 112	covalently tethered to 5'-terminus of DNA with flexible 9-carbon linker; intercalates between DNA bases; no global or local base stack distruptions extensive NMR and photophysical studies of DNA binding	E(Ru(III)/(II)) 1.6	ms
Et* 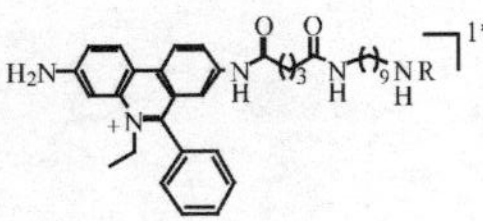	41, 96	covalently tethered to 5'-terminus of DNA with flexible 9-carbon linker; intercalates between DNA bases without stacking disruptions interaction of free Et with DNA exhaustively studied; numerous photophysical studies of tethered Et	E*(+/0) 1.2	2 ns
Ap* 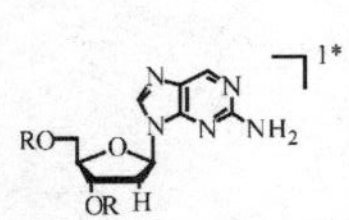	8, 40, 104, 105	incorporated as a base within DNA; base-pairing (T) and stacking very similar to A extensively characterized by a variety of photophysical studies; NMR structure	E*(0/-) 1.5	ps-ns
St* 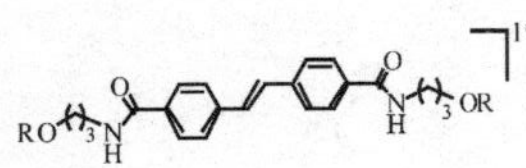	102, 169	connects two arms of a DNA hairpin; rigid; solvent-exposed 3.2 Å crystal structure of DNA hairpin connected by similar stilbene moiety	E*(0/-) 1.44	0.3-25 ns
Ap$^{\bullet+}$ 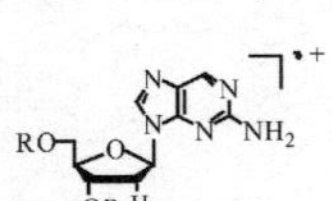	103, 170	incorporated as a base at the end of DNA; efficient formation of radical cation facilitated by fraying of the duplex ends T_m data on duplexes containing Ap at the terminus	E($^{\bullet+}$(-H)/0) 1.2	(•+) ~ 30 ns (•+(-H)) ms

Table 2 (continued)

Ac*	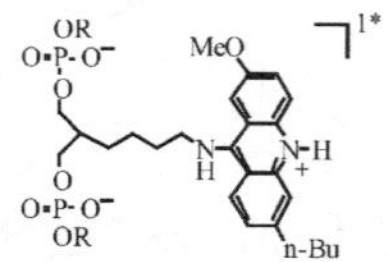	171-174	covalently attached to sugar-phosphate backbone; placed opposite A on complement; proposed to fill abasic pocket within duplex forcing the A to an extrahelical conformation T_m data; 2-D NMR; CD; fluorescence; propose rigid oxidant in highly-ordered environment	unknown	4-18 ns
AQ*	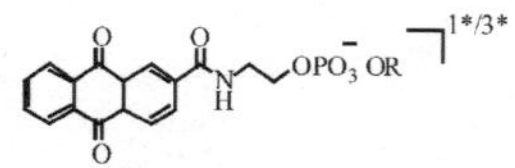	115	covalently linked to 5'-end of DNA by 2-carbon linker that prohibits intercalation; end-caps duplex; solvent-exposed T_m data; phosphorescence	E*(0/-) 2.2	1* < 10 ps
CNBP-U*	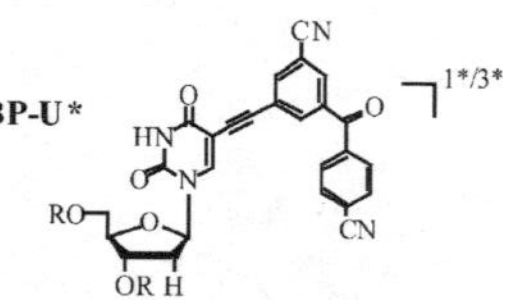	117-119	incorporated within DNA duplex T_m data	unknown	unknown
Sugar$^{\bullet+}$	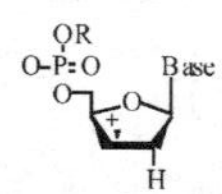	116	incorporated within DNA duplex; base = T, G; generation of radical cation induces strand break at position of oxidant	unknown	unknown

Oxidant/ Reductant	Evidence for CT	Injection Site/ Intervening Sequence	k_{ct}/Distance Regime	Distance Dependence	Sequence Specificity
$Rh(phi)_2(bpy')^{3+*}$/ GG $\Delta G < -0.7$ V	GGox PAGE 8-oxo-G product quenching by G in solution; detect G(-H) radical	Gs within 1-2 base-pairs of Rh(III) intercalator Gs in intervening sequence	GGox up to 200 Å	shallow GGox distal/proximal ~ 1 between 40 - 80 Å	strong GGox distal/proximal ~ 0.6 - 3.5 depending on intervening sequence
$Ru(dppz)(x)(y)^{3+}$/ GG $\Delta G < -0.3$ V	GG, M, tryptophan oxidation PAGE 8-oxo-G product	Gs within 1-2 base-pairs of Ru(III) intercalator	GGox up to 200 Å	shallow	strong
$Ru(dppz)(x)(y)^{3+}$/ M^c, tryptophan $\Delta G \sim -0.6$ V	detect G(-H), M, and tryptophan radicals by transient absorption, EPR	Gs in intervening sequence	k 10^7 s^{-1} up to 50 Å	CT not rate limiting up to 50 Å	GGox varies with sequence, base-stack perturbations; and extent of intercalation

Table 2 (continued)

Et*/^{Z}G $\Delta G \sim -0.2$ V	fluorescence quenching; correlates with CT driving force no spectral overlap for energy transfer	no Zs in injection site or intervening sequence	$k = 2 \times 10^{11}\ s^{-1}$ observed up to ~ 20 Å in time-resolved and steady-state experiments	shallow dependence of CT yield between 6-24 Å ($\gamma = 0.2$-0.3) no dependence of k_{CT} between 10-17 Å	strong efficiency and distance dependence of CT yield vary with Z flanking sequence and intervening mismatches
Ap*/G $\Delta G \sim -0.2$ V **Ap*/^{Z}G** $\Delta G \sim -0.5$ V	fluorescence quenching; correlates with CT driving force detection of G(-H) radical upon quenching of Ap* with dGTP in solution no spectral overlap for energy transfer	no Gs or Zs in injection site or intervening sequence	$k = 10^{11}$-$10^{9}\ s^{-1}$ observed up to ~ 14 Å in time-resolved and steady-state experiments	exponential distance dependence of CT rate constant and yield	strong k, efficiency and distance dependence of CT yield vary dramatically with intervening sequence, and differ for intra- versus interstrand processes
St*/G $\Delta G \sim 0.14$ V	fluorescence quenching stilbene radical anion detected in transient absorption	no Gs near injection site or in intervening sequence	$k = 10^{12}$-$10^{8}\ s^{-1}$ for 0-4 intervening A-T base-pairs (~3.4-17 Å)	exponential distance dependence of CT rate constant; $\beta \sim 0.6$-0.7 Å^{-1}	small variations in k depending on whether G is in the A or T arm of the hairpin
St*/G/GG $\Delta G \sim 0.14$ V	change in k_{decay} of stilbene radical anion monitored by transient absorption	Gs within 1-3 base-pairs of stilbene linker; CT to further GG sites	$k = 2 \times 10^{5}$-$8 \times 10^{7}\ s^{-1}$ 1-2 bases (3.4-6.8 Å)	strong order of magnitude decrease in k per intervening A	intrastrand CT faster through A than T bridges; intrastrand CT faster than interstrand CT
Ap$^{\bullet+}$/G **Ap$^{\bullet+}$(-H)/G(-H)** $\Delta G \sim 0$ V	detection of Ap(-H) and G(-H) radicals by transient absorption; detection of oxidation products by PAGE	no Gs near injection site or in intervening sequence	$k > 10^{7}\ s^{-1}$ for 0-6 intervening bases (~ 20 Å) $k > 10^{7}\ s^{-1}$ nearest neighbor; $k \sim 10^{6}$ - $< 10^{2}\ s^{-1}$ (1-6 intervening bases)	for T bridges, yield decreases ~ exponentially with distance for T bridges, exponential distance dependence of CT rate constant; $\beta \sim 0.75 \pm 0.2$ Å^{-1}	strong 10^{3} more efficient through A verus T bridges; no evidence for interstrand zig-zag mechanism; slower CT found to be proton-coupled
Ac*/G$_n$ (n = 1,2,3)	quenching of photoexcited acridine; formation of CT intermediate, and ground state recovery monitored by fluorescence and/or transient absorption	no Gs in injection site or intervening sequence	$k = 10^{11}$, 10^{8}-$10^{7}\ s^{-1}$ for nearest neighgor and one intervening base, respectively	anomalously large, implies $\beta > 2$ Å^{-1} propose increase in activation energy with increasing distance; cannot simply evaluate β	rate constants for nearest neighbor hole transfer, or through one A depend on the direction (5' vs 3')

Table 2 (continued)

AQ*/G_n (n = 2,3) ΔG < - 0.9 V	GGox PAGE anthraquinone radical anion observed by transient absorption	Gs within a few base-pairs of injection site and through intervening sequence	GGox up to ~ 200 Å	GGox distal/proximal ~ 0.5-0.05	strong GGox distal/proximal ~ 0.09 - 0.5 depending on sequence sequence near charge injection site modulates distal GGox
CNBP-U/G_n (n = 1,2,3)	GGox PAGE imidazolone detected as GGox product	Gs within a few base-pairs of injection site and through intervening sequence	examined up to 24 Å (7 intervening bases)	GGox distal/proximal ~ 0.8 - 1.6 for distal GG 3 base-pairs, and proximal GG 7 base pairs from oxidant	strong specific arrangement of G around injection site and intervening Gs required for CT
Sugar$^{\bullet+}$/G_n (n = 1,2,3)	GGox PAGE enol ether detected as product of reduction of sugar radical cation	Gs within a few base-pairs of injection site and through intervening sequence	charge injection: k = 10^9-10^6 s^{-1} for 0-3 intervening A-Ts (from yield ratios, trapping rate constant) charge transport: GGox up to 16 intervening base-pairs (~ 54 Å)	charge injection: exponential, β ~ 0.55 charge transport: 0-3 intervening bases, strong; 4-16 intervening bases, very shallow; propose change in mechanism	charge injection: strong, G or bases with lower potential are required near injection site charge transport: diminished CT through intervening mismatches

[a] x = phen, y= bpy'
[b] Please see other reviews in this text for a more detailed description of oxidants used to probe DNA-mediated charge transport.
[c] M = methyl indole

occupation of the DNA bridge is favored? This distance regime clearly also varies with donor, acceptor, and intervening bridge characteristics.

5.1 Charge Transfers Over a Short Range

Consider first hole transfer over relative short distances (<~20 Å), as probed between DNA bases (Ap^*, St^*, Ap radical cation), between photoexcited intercalators and DNA bases (Et^*, Ac^*), and between a terminally-bound photooxidant and bases within a DNA hairpin (St^*). Significantly, the measured rate constants for nearest-neighbor hole transfer (D-A distance ~3.4 Å) with low driving force are consistently 10^{11}–10^{12} s^{-1}. Rate constants for bridge-mediated hole transfer through one intervening base-pair (D-A distance ~6.8 Å) are generally between 10^{10}–10^{11} s^{-1} (Ap^*, St^*). A rate constant of 10^7–10^8 s^{-1} has been measured for hole transfer between guanines and an acridine photooxidant through an intervening A-T base-pair [171–174]. This somewhat anomalous behavior has been attributed to an increase in activation energy with distance [171]. The slow rate, may also, however, reflect poor coupling owing to a structural distortion within the assembly.

Slower bridge-mediated hole transfer has also been observed using photoionized Ap as the oxidant [103, 170]. Here two distinct hole transfer kinetics are observed; a fast reaction (>10^7 s^{-1}) attributed to direct oxidation of G by the Ap radical cation, and a slow reaction (10^6 s^{-1}) attributed to proton-coupled hole transfer between the deprotonated Ap radical cation and G. As the rate constant for the fast reaction is beyond the experimental time resolution, it may indeed be as fast as 10^{10}–10^{11} s^{-1}. Several key factors likely contribute to the apparently slow bridge-mediated hole transfer in these assemblies. First, the slower component monitors only hole transfer reactions which are slower than ~3×10^7 s^{-1}, the rate constant for oxidant formation, in this case for deprotonation of the Ap radical cation. Second, the base between Ap and G is a T; experiments in the same system at longer separations have found intervening As to mediate hole transfer with rate constants ~10^3-times larger [103]. Third, the distinctly different mechanism, namely proton-coupled hole transfer, may proceed with slower kinetics than direct hole transfer. Fourth, as Ap is placed at the duplex terminus, fraying of the duplex ends may reduce coupling between the oxidant and the DNA bases. Finally, the slow process is predicted to be essentially thermoneutral. Similarly slow rate constants of 10^8 s^{-1} have been measured for hole transfer between two guanine bases separated by a single A-T base pair [102].

The rate constants for these relatively short range hole transfer reactions generally decrease exponentially with distance. Yet, characterizing these DNA-mediated reactions with the parameter β is a simplification and is certainly inappropriate in cases where the Frank-Condon factor varies with distance (such as has been observed for the acridine photooxidant). Keeping these limitations in mind, however, β-values for DNA-mediated hole transfer of ~0.6–0.7 $Å^{-1}$ have been suggested using several different oxidant-DNA assemblies (Ap^*, St^*, Ap radical cation).

A distinctly different distance dependence has, however, been observed using Et* as an oxidant for hole transfer reactions with ^{Z}G [41, 96]. Here the rate constant for hole transfer, 2×10^{11} s^{-1}, is independent of distance between 10–17 Å. The yield of hole transfer falls off gradually over this distance regime. This behavior has been described in terms of hole transfer only for active conformations whose formation is gated by the molecular motions of the oxidant and/or DNA bases. Hole transfer within these optimally aligned D-B-A conformations exhibits a characteristically fast rate constant irrespective of distance, while the yield of hole transfer decreases with distance due to the reduced probability of achieving active conformations. The shallow distance dependence for this reaction was attributed to transport within the hopping regime. With photoexcitation of the donor, thermal occupation of bridging states is feasible.

Several factors may contribute to the fact that this behavior has been detected in Et-DNA, but not other oxidant-DNA assemblies. The fact that Et* bears a positive charge, while photoexcited stilbene, aminopurine, and the aminopurine radical, are neutral hole donors, may be a significant distinction, particularly in regulation of the reorganization energy. More important may be the interaction of Et with the DNA π-stack; Et is intimately coupled with the DNA bases; however it also exhibits significant and rapid molecular motions (τ~75 ps). The early result of efficient photoinduced quenching in the assembly containing metallointercalators as both donor and acceptor may represent another still better coupled example [95]. Indeed the combination of strong coupling with flexibility may be crucial for achieving conformations optimally active for hole transfer reactions. Assemblies with terminal oxidants including photoexcited stilbene and the aminopurine radical may lack sufficient flexibility, or sufficient stacking, respectively, for effective coupling with bridge states, thereby permitting transfer only through tunneling. Interestingly, in the case of CT involving photoexcited ethenoadenine and guanine, rates of CT are found to be remarkably slow (10^7 s^{-1}) with a steep distance dependence, and this has been attributed to very poor stacking of the base analogue [40]. It is also noteworthy in this context that for the stilbene hairpin assemblies, no hypochromism is observed in the stilbene chromophore, an indication of the absence of significant electronic overlap with neighboring DNA bases.

5.2 Long Range Charge Transport Through Multistep Hopping

Consider now longer-range hole transport reactions. Propagation of the hole over longer distances is usually achieved by a multistep process that involves occupation of the intervening DNA bridge. These long-range oxidations have been initiated by metallointercalators ($Rh(phi)_2(bpy')^{3+*}$, Ru(dppz) $(phen)(bpy')^{3+}$), end-capped anthraquinones (AQ*), a cyanobenzonphenone-modified uridine (CNBP-U*), and a DNA-sugar radical cation (sugar radical cation). In cases where the potentials are known ($Rh(phi)_2(bpy')^{3+*}$,

$Ru(dppz)(phen)(bpy')^{3+}$, AQ^{*}), and probably in the other cases as well, the oxidant lies energetically above the DNA bridge.

In all cases, the distance dependence of DNA-mediated hole transport has been seen to be relatively shallow. In the only case where the rate constant for a long-range reaction has been measured directly by following formation of the radical intermediate with transient absorption spectroscopy ($Ru(dppz)(phen)(bpy')^{3+}$), no distance-dependence was detected ($k>10^7$ s^{-1} up to 50 Å) [42, 108, 164]. But in this distance regime, generally, rates of CT have not been measured directly. Instead, the distance dependence in the yield of hole transport is typically evaluated from ratios of guanine oxidation at sites distal and proximal to the oxidant (d/p ratios). When considering hole traps of approximately equal potential, a d/p ratio of 1 indicates that the migrating hole equilibrates over the DNA duplex faster than it is trapped. Ratios that are greater than 1 suggest that the potential of the distal trap is deeper than the proximal trap, or that the rates of trapping are somehow different at the two sites. Ratios that are less than 1 indicate that the hole migration depends more strongly on distance, and complete equilibration does not occur. CT transport, rather than trapping is rate limiting.

But if these reactions all involve a multistep hopping scheme through the bridge, one might expect thermal relaxation at the first bridge state, after which point results should be equivalent irrespective of the oxidant. Nonetheless distinct variations are observed among different DNA assemblies. The metallointercalators ($Rh(phi)_2(bpy')^{3+*}$ and $Ru(dppz)(phen)(bpy')^{3+}$), consistently exhibit d/p ratios of ~1, or greater, over distances of 40–80 Å, indicative of rapid hole equilibration and a shallow distance dependence over these distances. Migration of the hole to longer distances (up to 200 Å) proceeds without a significant diminution in yield. Similarly, d/p ratios observed for HT initiated by a photoexcited cyanobenzophenone-modified uridine suggest no significant distance-dependence, although in this case the distance range does not extend beyond 24 Å. Moreover, the HT reaction is much more sensitive to the specific base-pair configuration of the injection site, perhaps a consequence of reduced coupling between the photooxidant and the DNA base stack. Conversely, HT initiated by an end-capped anthraquinone photooxidant exhibits a significantly steeper distance dependence; d/p ratios are typically 0.5–0.05. Certainly these data suggest that charge equilibration over the DNA duplex is not achieved following charge injection by the photoexcited anthraquinone. Here, the steeper distance dependence in the HT yield may be in part a consequence of the anthraquinone end-capped structure.

The extent of back electron transfer (BET) between the reduced oxidant and the guanine radical cation (radical) must also be considered when comparing the yields and distance dependence of oxidative damage induced by different photooxidants . We have recently obtained experimental evidence demonstrating the definitive role of BET. Using transient absorption spectroscopy we observed directly charge transfar between photoexited thionine (Th^{*}) and guanine [175], consistent with the highly favorable driving force for this reaction, and the previously reported quenching of intercalated Th^{*}

by G on the femtosecond time scale [176]. Yet, Th* fails to permanently oxidize guanine in duplex DNA. These observations indicate that trapping of the guanine radical cation (radical) is slow on the time scale of DNA-mediated CT; fast BET eliminates net CT chemistry. Indeed, by accelerating the rate of the guanine trapping using N^2-cyclopropylguanine, it is possible to permanently trap holes transferred from Th*, as well as Ap* which also does not permanently damage guanine in DNA [177]. Clearly, then the yield of oxidative damage need not reflect the rate constants for CT through DNA, and the rate of BET can modulate the extent and distribution of oxidative damage. Owing to differences between rhodium intercalators and anthraquinone in efficiencies of oxidative damage and in distal/proximal damage ratios, it was also suggested that results with the rhodium metallointercalcators might be artifactual, perhaps owing to aggregation [145, 178]. In our laboratory, under conditions used for oxidative damage experiments, we have examined and found no evidence for aggregation of assemblies containing tethered intercalators [179]. We have also directly compared efficiencies of photooxidation between capped anthraquinone and tethered rhodium and found them to be within the same order of magnitude. Importantly, we also examined photooxidation by theses tethered oxidants using the fast N^2-cyclopropylguanine trap. From this study [179], the discrepancies in damage yields and ratios among photooxidants can be understood based upon differing rates of BET among photooxidants. These results underscore the importance of distinguishing between rates and net yields of photooxidation in analyzing long range charge transport chemistry. Indeed, mechanistic descriptions of DNA-mediated CT and the resulting long range oxidative damage must include back electron transfer as a parameter.

Also difficult to reconcile, if there is thermal relaxation within the bridge on the timescale of the transport, is the ability to repair thymine dimers in DNA from a distant site oxidatively. While the oxidation potential of the thymine dimer dimer is not known, the oxidation potential of thymine is considered to be $\geq$1.8 eV, significantly above guanine (~1.2 eV). Yet efficient long-range thymine dimer repair has now been seen with several different photooxidants, all of which bind tightly to DNA by intercalation [149–151]. Does intercalation provide the requisite coupling for effective hole injection and rapid transport? Furthermore, the lack of competition seen between thymine dimer repair and oxidation of intervening guanine doublets argues that repair of the thymine dimer is faster than trapping at guanine sites [150]. Here, too, different rates of BET for different reduced oxidants may be a critical issue.

5.3
Electrochemistry: Is there Occupation of Bridge States?

In studies of photochemically induced CT, the energy of the donor and acceptor orbitals are close or higher in energy than the bridge states, and occupation of the bridge is understandable. In contrast, our electrochemical measurements employ redox active intercalators with reduction potentials

($E^{0/}$~−0.4–0.6 V vs NHE) which are significantly below the DNA bridge (for example, thymine $E^{0/-}$~−1.1 V vs NHE) [180]. Thermal access to the DNA bridge is improbable in these cases. It is noteworthy, however, that since the overall rate of electron transport is slow (10^2 s^{-1}), improbable thermal access may be feasible. In any case, electron transport over distances of 50–100 Å has been observed [93, 181], and it appears that what is rate-limiting is instead tunneling through the significantly shorter (15 Å) σ-bonded linker.

What mechanistic schemes can be invoked to rationalize electron transport through these DNA films? The rigidity imposed on the DNA duplexes within the tightly-packed films may provide some insight. For instance, this rigidity may facilitate the formation of well-coupled domains of π-stacked arrays. Alternatively, it has been proposed that electron transport through DNA films may be understood in terms of cooperativity effects among neighboring DNA duplexes and their surrounding ionic environment [182]. Electrochemical polarization of a well-ordered array of DNA duplexes is expected to be especially facile due to the mobility of positively-charged counterions associated with the negatively-charged phosphate backbones. As a consequence of this longitudinal polarizability, the DNA films are characterized by particularly large dielectric constants along the helical axis. Since, in semiconductors, the mean free path of an electron is inversely proportional to the dielectric constant, the polarized film is predicted to be highly conductive along the helical axis. In this context, disruption of electron transport by stacking perturbations such as mismatches can be rationalized in terms of defects within the polarized film.

5.4 Stacking Dynamics Control Charge Transport

Irrespective of the assembly examined, the importance of base stacking has emerged as the common denominator in all of our studies of DNA-mediated CT. Our results indicate that DNA CT can be rapid but only with effective coupling of donor and acceptor with the bridge, and effective coupling throughout the bridge due to aromatic π-stacking. Base stacking interactions are requisite for DNA-mediated CT, and variations in stacking can lead to significant variations in CT rates and efficiencies. Base stacking involves non-covalent interactions including solvation, electrostatic and dispersive forces; it is the dispersion interactions that are intimately related to the overlap area of the stacked bases. Stacking interactions facilitate electronic coupling among the DNA bases, and, correspondingly, CT through the DNA helix. What governs the strength of these interactions within DNA? The effective electronic coupling is determined by the individual coupling matrix elements between nearest neighbor bases, and between the donor and the bridge. However, it is the dynamic motions of the DNA bases, occurring on the picosecond to millisecond time scale, that ultimately define the magnitude and time dependence of this electronic coupling.

The significance of dynamic variations in stacking has been revealed by many of our investigations of DNA-mediated CT. The yield of long-range guanine oxidation through mismatched DNA was found to correlate most closely with dynamic motions at the mismatch site [142]. Conformational fluctuations of the DNA bases have likewise been shown to modulate the yield of CT, between two intercalators [64], between an intercalator and a DNA base [112], and between two DNA bases [105]. In time-resolved measurements of CT between Et* and $[Rh(phi)_2bpy']^{3+}$, it was the yield, rather than the rate constant, which decreased with increasing donor-acceptor distance [64]. Studies of DNA-mediated CT between Et* and ^{Z}G, including measurements of CT dynamics with picosecond time resolution, likewise observed rate constants which were largely insensitive to distance (up to 17 Å examined), while the yields diminished with increasing donor-acceptor separation [41, 96]. Furthermore, two components were detected in the decay of Et*, a fast component (5 ps) corresponding to direct CT with ^{Z}G, and a slow component (75 ps) corresponding to motion of Et* at its intercalation site. We have also now seen a remarkable variation in quenching of Ap* as a function of temperature [183, 184]. CT over a longer distance (at least 30 Å) is made possible with increasing temperature, as if the higher temperatures allow greater dynamical access to conformations favorable for CT. Hole propogation out to longer distance was also evident with increasing temperature in our studies of long-range oxidative DNA damage [112]. In all cases, however, once temperatures are reached where the unstacking of the duplex occurrs, CT is lost.

These observations, and others made in the course of our investigations, can be interpreted in terms of a model where the conformational dynamics of DNA regulate CT [184]. In this model, only certain, well-coupled, conformations are active towards CT. Rapid CT with a shallow distance dependence occurs within these optimal conformations; other conformations do not achieve appropriate coupling in sufficient time (such as the excited state lifetime of the donor) and do not undergo CT, or exhibit slower CT dynamics corresponding to the timescale for rate-limiting reorganizational motions. In this way it is possible for such motions to act as a gate for CT reactions in DNA. The yield of CT is correlated with the probability of a CT event; that is, the probability that a CT-active conformation is achieved. This probability generally decreases with increasing donor-acceptor distance, as the number of bases that must be appropriately aligned increases.

How do we envision these CT-active conformations? Certainly they will correspond to arrangements in which the effective electronic coupling is maximized. Indeed, these active conformations may represent domains of transient charge delocalization. Importantly, the size and character of such domains will be defined by sequence-dependent structure and dynamics. While it is unlikely that a domain will extend over distances of 200 Å, even transiently, migration of charges over long molecular distances can be described as motion between these domains. Here the dynamic flexibility of the DNA bases propagates the charge, facilitating transport between these delocalized domains.

This model provides a framework for reconciling many of the different experiments carried out to probe DNA CT. Models that describe DNA CT simply in terms of donor, acceptor, and bridge energetics and distances are insufficient to account for the many variations we have seen. DNA CT also depends sensitively upon the extensive couplings within the base pair stack, and these are defined by the sequence-dependent structure and dynamics of DNA. Indeed, DNA CT chemistry may provide a sensitive means to monitor and describe these sequence-dependent dynamical motions.

6
Conclusions

The analogy drawn between π-stacked solids and duplex DNA has provided a useful starting point for experiments to probe and understand DNA-mediated CT. As with the π-stacked solids, the DNA base pair array can provide an effective medium for long range CT. Mechanistically, however, the differences between DNA and these solid state materials may be even more important to consider. Duplex DNA, as a molecular π-stacked structure, undergoes dynamical motion in solution. The time-dependent and sequence-dependent structures that arise serve to modulate and gate CT. Indeed in probing DNA CT as a function of sequence and sequence-dependent structure, we may better understand mechanistically how CT proceeds and how DNA CT may be utilized.

DNA CT also permits chemistry at a distance. Oxidative DNA damage and thymine dimer repair can proceed in a DNA-mediated reaction initiated from a remote site. These reactions too are sensitive to intervening DNA dynamical structure, and such structures can serve to modulate DNA CT chemistry. The sensitivity of DNA CT to base pair stacking also provides the basis for the design of new DNA diagnostics, tools to detect mutations in DNA and to probe protein-DNA interactions.

But is DNA CT chemistry also exploited within the cell? Has nature devised schemes to protect the genome from oxidative damage from a distance? Has nature also exploited CT chemistry in detecting damage from a distance, or more generally, in devising strategies for long range DNA-mediated signaling? These are challenges for the future, but surely the DNA double helix will provide still more surprises and rich chemistry to discover.

References

1. Watson JD, Crick FHC (1953) Nature 171:737
2. Szent-Györgi A (1960) Proc Natl Acad Sci USA 46:1444
3. Eley DD, Spivey DI (1962) Trans Faraday Soc 58:411
4. Miller JS, Epstein AJ (1994) Angew Chem Int Edit 33:385
5. Fox MA (1992) Acc Chem Res 25:569
6. Marks TJ (1985) Science 227:881
7. Saenger W, Cantor CR (ed) (1984) Principles of Nucleic Acid Structure. Springer, Berlin Heidelberg New York
8. Nordlund TM, Andersson S, Nilsson L, Rigler R, Gräslund A, McLaughlin LW (1989) Biochemistry 28:9095
9. Guest CR, Hochstrasser RA, Sowers LC, Millar DP (1991) Biochemistry 30:3271

10. Brauns EB, Madaras ML, Coleman RS, Murphy CJ, Berg MA (1999) J Am Chem Soc 121:11644
11. Brauns EB, Murphy CJ, Berg MA (1998) J Am Chem Soc 120:2449
12. Liang Z, Freed JH, Keyes RS, Bobst AM (2000) J Phys Chem B 104:5372
13. Georghiou S, Bradrick TD, Philippetis A, Beechem JM (1996) Biophys J 70:1909
14. Borer PN, LaPlante SR, Kumar A, Zanatta N, Martin A, Hakkinen A, Levy GC (1994) Biochemistry 33:2441
15. Cheatham III TE, Kollman PA (2000) Biopolymers 56:232
16. Giudice E, Lavery R (2002) Acc Chem Res 35:350
17. Snart RS (1963) Trans Faraday Soc 59:754
18. Snart RS (1968) Biopolymers 6:293
19. Liang CY, Scalco EG (1964) J Chem Phys 40:919
20. Warman JM, de Haas MP, Rupprecht A (1996) Chem Phys Lett 249:319
21. Okahata Y, Kobayashi T, Tanaka K, Shimomura M (1998) J Am Chem Soc 120:6165
22. Fink HW, Schönenberger C (1999) Nature 398:407
23. Porath D, Bezryadin A, de Vries S, Dekker C (2000) Nature 403:635
24. Dekker C, Ratner MA (2001) Physics World 29:33
25. Kasumov AY, Kociak M, Guéron S, Reulet B, Volkov VT, Klinov DV, Bouchiat H (2001) Science 291:280
26. Gray HB, Winkler JR (1996) Annu Rev Biochem 65:537
27. O'Neill P, Fielden EM (1993) Adv Radiat Biol 17:53
28. Barton JK, Kumar CV, Turro NJ (1986) J Am Chem Soc 108:6391
29. Purugganan MD, Kumar CV, Turro NJ, Barton JK (1988) Science 241:1645
30. Murphy CJ, Arkin MR, Ghatlia ND, Bossmann S, Turro NJ, Barton JK (1994) Proc Natl Acad Sci USA 91:5315
31. Arkin MR, Stemp EDA, Holmlin RE, Barton JK, Hormann A, Olson EJC, Barbara PF (1996) Science 273:475
32. Arkin MR, Stemp EDA, Turro C, Turro NJ, Barton JK (1996) J Am Chem Soc 118:2267
33. Stemp EDA, Arkin MR, Barton JK (1995) J Am Chem Soc 117:2375
34. Holmlin RE, Stemp EDA, Barton JK (1996) J Am Chem Soc 118:5236
35. Olson EJC, Hu D, Hörmann A, Barbara PF (1997) J Phys Chem 101:299
36. Lincoln P, Tuite E, Nordén B (1997) J Am Chem Soc 119:1454
37. Brun AM, Harriman A (1994) J Am Chem Soc 116:10383
38. Meade TJ, Kayyem JF (1995) Angew Chem Int Edit 34:352
39. Steenken S, Jovanovic SV (1997) J Am Chem Soc 119:617
40. Kelley SO, Barton JK (1999) Science 283:375
41. Kelley SO, Barton JK (1998) Chem Biol 5:413
42. Pascaly M, Yoo J, Barton JK (2002) J Am Chem Soc 124:9083
43. Turro C, Evenzahav A, Bossmann SH, Barton JK, Turro NJ (1996) Inorg Chim Acta 243:101
44. Núñez ME, Barton JK (2000) Curr Opin Chem Biol 4:199
45. Erkkila KE, Odom DT, Barton JK (1999) Chem Rev 99:2777
46. Holmlin RE, Dandliker PJ, Barton JK (1999) Bioconjugate Chem 10:1122
47. Sitlani A, Long EC, Pyle AM, Barton JK (1992) J Am Chem Soc 114:2303
48. Sitlani A, Barton JK (1994) Biochemistry 33:12100
49. David SS, Barton JK (1998) J Am Chem Soc 115:2984
50. Hudson BP, Barton JK (1998) J Am Chem Soc 120:6877
51. Kielkopf CL, Erkkila KE, Hudson BP, Barton JK, Rees DC (2000) Nat Struct Biol 7:117
52. Turro C, Hall DB, Chen W, Zuilhof H, Barton JK, Turro NJ (1998) J Phys Chem A 102:5708
53. Friedman AE, Chambron JC, Sauvage JP, Turro NJ, Barton JK (1990) J Am Chem Soc 112:4960
54. Jenkins Y, Friedman AE, Turro NJ, Barton JK (1992) Biochemistry 31:10809
55. Olson EJC, Hu D, Hörmann A, Jonkman AM, Arkin MR, Stemp EDA, Barton JK, Barbara PF (1997) J Am Chem Soc 119:11458

56. Dupureur CM, Barton JK (1994) J Am Chem Soc 116:10286
57. Haq I, Lincoln P, Suh D, Nordén B, Chowdry BZ, Chaires JB (1995) J Am Chem Soc 117:4788
58. Stemp EDA, Holmlin RE, Barton JK (2000) Inorg Chim Acta 297:88
59. Stemp EDA, Arkin MR, Barton JK (1997) J Am Chem Soc 119:2921
60. Chang IJ, Gray HB, Winkler JR (1991) J Am Chem Soc 113:7056
61. Waring MJ (1965) J Mol Biol 13:269
62. Lepecq JB, Paoletti C (1967) J Mol Biol 27:87
63. Olmsted J, Kearns DR (1977) Biochemistry 16:3647
64. Kelley SO, Holmlin RE, Stemp EDA, Barton JK (1997) J Am Chem Soc 119:9861
65. Bradley DF, Stellwagen NC, O'Konski CT, Paulson CM (1972) Biopolymers 11:645
66. Nordén B, Tjerneld F (1982) Biopolymers 21:1713
67. Tuite E, Nordén B (1994) J Am Chem Soc 116:7548
68. Tuite E, Kelly JM (1995) Biopolymers 35:419
69. Arcamone F (1981) Doxorubicin: anticancer antibiotics. Academic Press, New York
70. Leng F, Savkur R, Fokt I, Przewloka T, Priebe W, Chaires JB (1996) J Am Chem Soc 118: 4732
71. Wang AHJ, Gao YG, Liaw YC, Li YK (1991) Biochemistry 30:3812
72. Pecourt JML, Peon J, Kohler B (2000) J Am Chem Soc 122:9348
73. Onidas D, Markovitsi D, Marguet S, Sharonov A, Gustavsson T (2002) J Phys Chem B 106:11367
74. Barrio JR, Secrist JA, Leonard NJ (1972) Biochem Biophys Res Commun 46:597
75. Secrist JA, Barrio JR, Leonard NJ, Weber G (1972) Biochemistry 11:3499
76. Spencer RD, Weber G, Tolman GL, Barrio JR, Leonard NJ (1974) Eur J Biochem 45:425
77. Ward DC, Reich E, Stryer L (1969) J Biol Chem 244:1228
78. Nir E, Kleinermanns K, Grace L, de Vreis MS (2001) J Phys Chem A 105:5106
79. Xu DG, Evans KO, Nordlund TM (1994) Biochemistry 33:9592
80. Hochstrasser RA, Carver TE, Sowers LC, Millar DP (1994) Biochemistry 33:11971
81. Allan WB, Reich NO (1996) Biochemistry 35:14757
82. Holz B, Klimasauskas S, Serva S, Weinhold E (1998) Nucleic Acids Res 26:1076
83. Rachofsky EL, Seibert E, Stivers JT, Osman R, Ross JBA (2001) Biochemistry 40:957
84. Mandal SS, Fidalgo da Silva E, Reha-Krantz LJ (2002) Biochemistry 41:4399
85. Kouchakdjian M, Eisenberg M, Yarema K, Basu A, Essigmann J, Patel DJ (1991) Biochemistry 30:1820
86. Nakatani K, Dohno C, Saito I (2000) J Am Chem Soc 122:5893
87. Voityuk AA, Rösch N (2002) J Phys Chem B 106:3013
88. Kool ET, Morales JC, Guckian KM (2000) Angew Chem Int Edit 39:990
89. Prütz WA, Land EJ (1979) Int J Radiat Biol 36:513
90. Sugiyama H, Saito I (1996) J Am Chem Soc 118:7063
91. Burrows CJ, Muller JG (1998) Chem Rev 98:1109
92. Kelley SO, Barton JK, Jackson NM, Hill MG (1997) Bioconjugate Chem 8:31
93. Kelley SO, Jackson NM, Hill MG, Barton JK (1999) Angew Chem Int Edit 38:941
94. Napper AM, Liu H, Waldeck DH (2001) J Phys Chem B 105:7699
95. Murphy CJ, Arkin MR, Jenkins Y, Ghatlia ND, Bossmann SH, Turro NJ, Barton JK (1993) Science 262:1025
96. Wan CZ, Fiebig T, Kelley SO, Treadway CR, Barton JK, Zewail AH (1999) Proc Natl Acad Sci USA 96:6014
97. Fiebig T, Wan CZ, Kelley SO, Barton JK, Zewail AH (1999) Proc Natl Acad Sci USA 96:1187
98. Davis WB, Ratner MA, Wasielewski MR (2001) J Am Chem Soc 123:7877
99. Berlin YA, Burin AL, Siebbeles LDA, Ratner MA (2001) J Phys Chem A 105:5666
100. Grozema FC, Siebbeles LDA, Berlin YA, Ratner MA (2002) Chem Phys Chem 6:536
101. Wan CZ, Fiebig T, Schiemann O, Barton JK, Zewail AH (2000) Proc Natl Acad Sci USA 97:14052
102. Lewis FD, Zuo X, Liu J, Hayes RT, Wasielewski MR (2002) J Am Chem Soc 124:4568

103. Shafirovich V, Dourandin A, Huang WD, Luneva NP, Geacintov NE (2000) Phys Chem Chem Phys 2:4399
104. O'Neill MA, Barton JK (2002) Proc Natl Acad Sci USA 99:16543
105. O'Neill MA, Barton JK (2002) J Am Chem Soc 124:13053
106. Candeias LP, Steenken S (1989) J Am Chem Soc 111:094
107. Schiemann O, Turro NJ, Barton JK (2000) J Phys Chem B 104:7214
108. Yoo J, Delaney S, Stemp EDA, Barton JK (2003) J Am Chem Soc 125:6640
109. Hall DB, Holmlin RE, Barton JK (1996) Nature 382:731
110. Hall DB, Barton JK (1997) J Am Chem Soc 119:5045
111. Prat F, Houk KN, Foote CS (1998) J Am Chem Soc 120:845
112. Núñez ME, Hall DB, Barton JK (1999) Chem Biol 6:85
113. Arkin MR, Stemp EDA, Pulver SC, Barton JK (1997) Chem Biol 4:389
114. Hall DB, Kelley SO, Barton JK (1998) Biochemistry 37:15933
115. Schuster GB (2000) Accounts Chem Res 33:253
116. Giese B (2002) Annu Rev Biochem 71:51
117 Nakatani K, Dohno C, Saito I (1999) J Am Chem Soc 121:10854
118. Nakatani K, Dohno C, Saito I (2000) Tetrahedron Lett 41:10041
119. Nakatani K, Dohno C, Saito I (2001) J Am Chem Soc 123:9681
120. Delaney S, Pascaly M, Bhattacharya PK, Han K, Barton JK (2002) Inorg Chem 41:1966
121. Henderson PT, Jones D, Hampikian G, Kan YZ, Schuster GB (1999) Proc Natl Acad Sci USA 96:8353
122. Hartmann B, Lavery R (1996) Q Rev Biophys 29:309
123. Belmont P, Constant JF, Demeunynck (2001) Chem Soc Rev 30:70
124. Kan YZ, Schuster GB (1999) J Am Chem Soc 121:10857
125. Kan YZ, Schuster GB (1999) J Am Chem Soc 121:11607
126, Núñez ME, Noyes KT, Gianolio DA, McLaughlin LW, Barton JK (2000) Biochemistry 39:6190
127. Odom DT, Dill EA, Barton JK (2001) Nucleic Acids Res 29:2026
128. Odom DT, Dill EA, Barton JK (2000) Chem Biol 7:475–481
129. Sartor V, Henderson PT, Schuster GB (1999) J Am Chem Soc 121:11027
130. Odom DT, Barton JK (2001) Biochemistry 40:8727
131. Abdou IM, Sartor V, Cao HC, Schuster GB (2001) J Am Chem Soc 123:6696
132. Williams TT, Odom DT, Barton JK (2000) J Am Chem Soc 122:9048
133. Meggers E, Michel-Beyerle ME, Giese B (1998) J Am Chem Soc 120:12950
134. Bixon M, Jortner J (2000) J Phys Chem B 104:3906
135. Jortner J, Bixon M, Langenbacher T, Michel-Beyerle ME (1998) Proc Natl Acad Sci USA 95:12759
136. Nadeau JG, Crothers DM (1989) Proc Natl Acad Sci USA 86:2622
137. Berlin YA, Burin AL, Ratner MA (2002) Chem Phys 275:61
138. Dickerson RE (1998) Nucleic Acids Res 26:1906
139. Brown T, Hunter WN, Kneale G, Kennard O (1986) Proc Natl Acad Sci USA 83 2402
140. Hunter WN, Brown T, Kennard O (1987) Nucleic Acids Res 15:6589
141. Allawi HT, SantaLucia J Jr (1998) Nucleic Acids Res 26:4925
142. Bhattacharya PK, Barton JK (2001) J Am Chem Soc 123:8649
143. Giese B, Wessely S (2000) Angew Chem Int Edit 39:3490
144. Williams TT, Barton JK (2002) J Am Chem Soc 124:1840
145. Santhosh U, Schuster GB (2002) J Am Chem Soc 124:10986
146. Sancar A (1996) Annu Rev Biochem 65:43
147. Jacobsen JR, Cochran AG, Stephans JC, King DS, Schultz PG (1995) J Am Chem Soc 117:5453
148. Young T, Nieman R, Rose SD (1990) Photochem Photobiol 52:661
149. Dandliker PJ, Holmlin RE, Barton JK (1997) Science 275:1465
150. Dandliker PJ, Núñez ME, Barton JK (1998) Biochemistry 37:6491
151. Vicic DA, Odom DT, Núñez ME, Gianolio DA, McLaughlin LW, Barton JK (2000) J Am Chem Soc 122:8603

152. Dotse AK, Boone EK, Schuster GB (2000) J Am Chem Soc 122:6825
153. Kelley SO, Barton JK, Jackson NM, McPherson LD, Potter AB, Spain EM, Allen MJ, Hill MG (1998) Langmuir 14:6781
154. Sam M, Boon EM, Barton JK, Hill MG, Spain EM (2001) Langmuir 17:5727
155. Kelley SO, Boon EM, Barton JK, Jackson NM, Hill MG (1999) Nucleic Acids Res 27:4830
156. Boon EM, Ceres DM, Drummond TG, Hill MG, Barton JK (2000) Nat Biotechnol 18:1096
157. Boon EM, Barton JK, Pradeepkumar PI, Isaksson J, Petit C, Chattopadhyaya J (2002) Angew Chem Int Edit 41:3402
158. Boon EM, Salas JE, Barton JK (2002) Nat Biotechnol 20:282
159. Rajski SR, Kumar S, Roberts RJ, Barton JK (1999) J Am Chem Soc 121:5615
160. Rajski SR, Barton JK (2001) Biochemistry 40:5556
161. Boon EM, Pope MA, Williams SD, David SS, Barton JK (2002) Biochemistry 41:8464
162. Wagenknecht HA, Stemp EDA, Barton JK (2000) Biochemistry 39:5483
163. Wagenknecht HA, Stemp EDA, Barton JK (2000) J Am Chem Soc 122:1
164. Wagenknecht HA, Rajski SR, Pascaly M, Stemp EDA, Barton JK (2001) J Am Chem Soc 123:4400
165. Stemp EDA, Barton JK (2000) Inorg Chem 39:3868
166. Boon EM, Livingston AL, Chmiel NA, David SS, Barton JK (2003) Proc Natl Acad Sci USA 100:12543
167. Núñez ME, Noyes KT, Barton JK (2002) Chem Biol 9:403
168. Núñez ME, Holmquist GP, Barton JK (2001) Biochemistry 40:12465
169. Lewis FD, Letsinger RL, Wasielewski MR (2001) Accounts Chem Res 34:159
170. Shafirovich V, Dourandin A, Geacintov NE (2001) J Phys Chem B 105:8431
171. Hess S, Gotz M, Davis WB, Michel-Beyerle ME (2001) J Am Chem Soc 123:10046
172. Davis WB, Naydenova I, Haselsberger R, Ogrodnik A, Giese B, Michel-Beyerle ME (2000) Angew Chem Int Edit 39:3649
173. Fukui K, Tanaka K, Fujitsuka M, Watanabe A, Ito O (1999) J Photochem Photobiol B 50:18
174. Fukui K, Tanaka K (1998) Angew Chem Int Edit 37:158
175. Dohno C, Stemp EDA, Barton JK (2003) J Am Chem Soc 125:9586
176. Reid GD, Whittaker DJ, Day MA, Turton DA, Kayser V, Kelly JM, Beddard GS (2002) J Am Chem Soc 124:5518
177. O'Neill MA, Dohno C, Barton JK (2003) (submitted)
178. Fahlman RP, Sharma RD, Sen D (2002) J Am Chem Soc 124:12477
179. Williams T, Dohno C, Stemp EDA, Barton JK (2003) (unpublished results in our laboratory)
180. Steenken S, Telo JP, Novais HM, Candeias LP (1992) J Am Chem Soc 114:4701
181. Liu T, Barton JK (2002) (unpublished results in our laboratory)
182. Hartwich G, Caruana DJ, de Lumley-Woodyear T, Wu YB, Campbell CN, Heller A (1999) J Am Chem Soc 121:10803
183. O'Neill MA, Barton JK (2002) (unpublished results in our laboratory)
184. O'Neill MA, Beeker HC, Wan C, Barton JK, Zewail AH (2003) Angew Chem Int Ed 42: in press

Top Curr Chem (2004) 236:117–137
DOI 10.1007/b94413

Hole Transfer in DNA by Monitoring the Transient Absorption of Radical Cations of Organic Molecules Conjugated to DNA

Kiyohiko Kawai · Tetsuro Majima

The Institute of Scientific and Industrial Research (SANKEN), Osaka University, Mihogaoka 8-1, Ibaraki, 567-0047 Osaka, Japan
E-mail: majima@sanken.osaka-u.ac.jp

Abstract Hole transfer in DNA has attracted much interest, since the hole causes oxidative damage in vivo which leads to mutation. Furthermore, the occurrence of long-range hole transfer in DNA suggests a potential application of DNA as a building block for biosensors and bioelectronic devices. DNA is an attractive molecule to connect two points selected according to the information of DNA carried as the base sequence. In addition, DNA can also be used to construct highly-ordered structures in nanoscale devices. Therefore hole transfer in DNA and factors controlling its rates have been extensively studied.

For an understanding of the kinetics of hole transfer in DNA, time-resolved spectroscopic measurements are required. However, there are few reports that provide direct evidence for the formation and decay of the radical cation intermediate. In this chapter, we used a pulse radiolysis technique to investigate the hole transfer process in DNA. Pulse radiolysis provides a unique system for the observation of charge transfer processes over 1 μs, including the multi-step hole transfer process. For direct observation of the radical cation species during the hole transfer process in DNA, we synthesized ODNs site-specifically conjugated with special probe molecules, of which the radical cation has a distinct absorption peak with a large molar extinction coefficient. Using this pulse radiolytic study of synthetic ODNs, the kinetics of the hole transfer in DNA were investigated.

Keywords Pulse radiolysis · Pyrene · Phenothiazine · Radical cation · Hole transfer rate

Abbreviations

ODN	Oligodeoxynucleotide
G	Guanine
A	Adenine
C	Cytosine
T	Thymine
oxG	7,8-Dihydro-8-oxo-deoxyguanine
Py	Pyrene
Ptz	Phenothiazine
2AP	2-Amionpurine
$^{\cdot+}$	Radical cation
NHE	Normal hydrogen electrode
T_m	Melting temperature

1 Introduction

1.1 Hole Transfer in DNA

One-electron oxidation of DNA leads to formation of the radical cation of guanine ($G^{\cdot+}$), the nucleobase with the lowest oxidation potential (G:1.47>A:1.94>C:2.1~T:2.1, V vs NHE in CH_3CN) [1]. It is well established that a hole migrates through DNA by hopping between Gs. Indeed, the occurrence of multi-step long-range hole transfer over a distance of 200 Å has been demonstrated by Giese [2, 3], Schuster [4, 5], and Barton [6]. Long-range hole transfer was evidenced by strand cleavage analysis using photochemical reactions on DNA assemblies containing tethered photooxidant. After the generation of $G^{\cdot+}$ near the photooxidant, the hole migrated over a long range through DNA by multi-step hopping between Gs, and oxidative strand scission occurs far away from the initial oxidation site. The hole transfer in DNA and the factors controlling the rates have been extensively studied since the hole may cause oxidative damage in vivo, which leads to mutation and strand scission [7, 8].

On the other hand, if the hole flow in DNA could be artificially controlled to deposit at the desired site in DNA, it may enable site-selective oxidation and strand scission of DNA, which is desirable from a therapeutical standpoint. Furthermore, understanding DNA-mediated hole transfer is expected to lead to an additional application in the development of biosensors and bioelectronic devices [9]. Therefore, the regulation of the transfer rate and direction of the hole generated in DNA is of interest from the perspective of using DNA as a building block for electronic devices.

1.2
Pulse Radiolysis Measurements of Hole Transfer Rate

Time-resolved spectroscopic measurements are needed to clarify the kinetics of the hole transfer in DNA; photochemical methods have usually been used to study this charge transfer process. During initial studies of the charge transfer dynamics in DNA, the electron transfer rate in DNA was been measured by luminescence quenching experiments [10–14]. A fluorescent chromophore was conjugated at the defined site of oligodeoxynucleotide (ODN) as a photooxidant, and the rates of the electron transfer from the nearby G were measured by time-resolved fluorescence spectroscopy. Electron transfer rates in the time range 10^5–10^{12} s^{-1} have been reported to be dependent upon the distance between the electron donor and acceptor, nature of fluorescent chromophores, and the free energy of the electron transfer process in each system. After the photo-initiated charge separation process, the hole transfer may occur from the generated $G^{\cdot+}$ to the second hole acceptor, such as G, GG, or an artificial base with lower oxidation potential than that of G [2, 15, 16]. Lewis et al have successfully determined the single-step hole transfer rate constants from $G^{\cdot+}$ to GG as 5×10^7 s^{-1} by monitoring the decay of the stilbenedicarboxamide radical anion [17]. For the multi-step hole transfer in DNA to occur using the photochemical method, the radical anion of the electron acceptor which accompanies the generation of $G^{\cdot+}$ should be consumed to avoid charge recombination diminishing $G^{\cdot+}$. This can be achieved, for example, by the reaction of radical anion of a photosensitizer with molecular oxygen. The concentration of molecular oxygen in water is approximately 0.2 mM at room temperature. Since the diffusion-controlled rate constant in water at room temperature is approximately 7×10^9 M^{-1} s^{-1}, the upper limit of the rate of this reaction is 10^6 s^{-1}. Hence, the charge recombination process is usually much more efficient, and the quantum yield of hole injection is often very low. Therefore photo-induced electron transfer is not suited for the measurement of hole transfer rate in DNA slower than 10^5 s^{-1}, and hole transfer including the multi-step hole transfer process following the fast charge recombination between the radical anion of the photosensitizer and $G^{\cdot+}$. For the measurement of hole transfer rate slower than 10^5 s^{-1}, it is essential to generate the radical cation (hole) selectively in DNA, without the generation of radical anion. For this purpose, pulse radiolysis serves as a suitable method.

Pulse radiolysis has been widely used as a convenient method to measure the charge transfer rate in various intra- and intermolecular systems [18–20]. Pulse radiolysis and laser photochemical techniques are truly complementary and great progress in understanding the charge transfer process has come from the application of both techniques. Pulse radiolysis has a particular advantage with the production of radical ions, since the desired radical ion is usually the only species that absorbs light in the visible and near-infrared regions. Electron pulses with a high energy can be used to generate as much as 10 μM of the radical ions of the solute molecules. Therefore, pulse radiolysis has been widely used to determine the spectroscopic properties of the radical ions of various molecules [21–33]. The absorption properties of radical ion species of nucleobases were also determined using pulse radiolysis by Steenken et al [34–38]. Since the charge can be irreversibly generated in the medium of interest and is free from charge recombination, pulse radiolysis provides a unique system for the observation of charge transfer processes over 1 μs.

Pulse radiolysis of dilute aqueous solutions of DNA mainly causes ionization of the most abundant molecules, water, producing hydrated electrons (e_{aq}^-), hydroxyl radicals ($^{\cdot}OH$), and hydrogen atoms ($^{\cdot}H$) as the primary reactive species, with *G*-values of 2.6, 2.7, and 0.6, respectively [39], where the *G*-value defines the number of entities formed as a result of expenditure of radiation energy of 100 eV. Using an 8 ns electron pulse from a 28-MeV linear accelerator (LINAC) at SANKEN, Osaka University, these reactive species are generated at a concentration of 10 μM. In the presence of 10 mM $S_2O_8^{2-}$ and 100 mM *t*-BuOH, e_{aq}^- reacts with $S_2O_8^{2-}$ to give the strong one-electron oxidant $SO_4^{\cdot-}$, whereas the undesired $^{\cdot}OH$ is scavenged from solution by the hydrogen-abstraction reaction from *t*-BuOH, to give a relatively stable radical $^{\cdot}CH_2(CH_3)COH$ [34] (Scheme 1). Otherwise, under N_2O-saturated

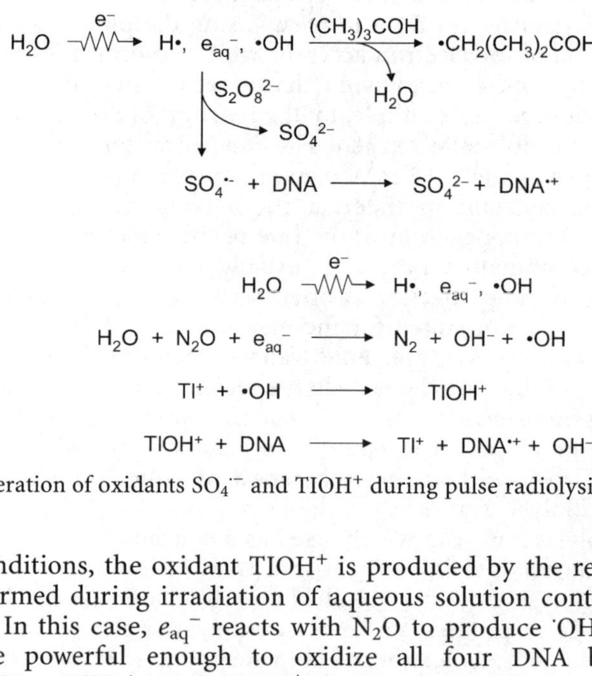

Scheme 1 Generation of oxidants $SO_4^{\cdot-}$ and $TlOH^+$ during pulse radiolysis

(22 mM) conditions, the oxidant $TlOH^+$ is produced by the reaction of Tl^+ with $^{\cdot}OH$, formed during irradiation of aqueous solution containing 1 mM Tl_2SO_4 [40]. In this case, e_{aq}^- reacts with N_2O to produce $^{\cdot}OH$. Since these oxidants are powerful enough to oxidize all four DNA bases ($SO_4^{\cdot-}$: E_{ox}=2.5~3.1 V vs NHE in H_2O, $TlOH^+$: E_{ox}=2.2 V vs NHE in H_2O), random oxidation of the whole DNA strand is expected. These oxidants are used in the next chapter to generate a hole in DNA.

1.3 Synthesis of Probe-Conjugated ODN

Even after a hole is successfully generated in DNA at a high concentration, observation of the hole transfer by monitoring the transient absorption of $G^{\cdot+}$ is difficult due to the low extinction coefficient of $G^{\cdot+}$ [35] (less than 3000 M^{-1} cm^{-1} over 350 nm). Furthermore, it is hard to define the site of $G^{\cdot+}$ since DNA consists of only four bases, resulting in many Gs within a modest length of ODN. For the observation of the radical cation generated in DNA, we synthesized ODNs site-specifically conjugated with probe molecules. The radical cation of probe molecules is required to have a distinct absorption peak with a large molar extinction coefficient, and also needs to be relatively chemically inert: that is to have long lifetime in aqueous solution.

According to these requirements, we selected pyrene (Py) and phenothiazine (Ptz) [41] as the probe molecules in DNA. Radical cations of Py and Ptz ($Py^{\cdot+}$ and $Ptz^{\cdot+}$) have strong absorption peaks in the visible region (λ_{max}=470 nm and 520 nm, $\varepsilon_{max}\approx2\times10^4$ and 9×10^3 M^{-1} cm^{-1}, respectively). Site-specifically Py- and Ptz-conjugated ODNs were synthesized by the automated solid-phase phosphoramidite method according to the reported procedure. In section 2.1, 5′-Py- or 5′-Ptz-conjugated ODNs were synthesized [42–44] in order to observe the multi-step hole transfer from $G^{\cdot+}$ generated in DNA to Py or Ptz (Scheme 2) [45, 46]. In section 2.2, 5′-PyODNs contain-

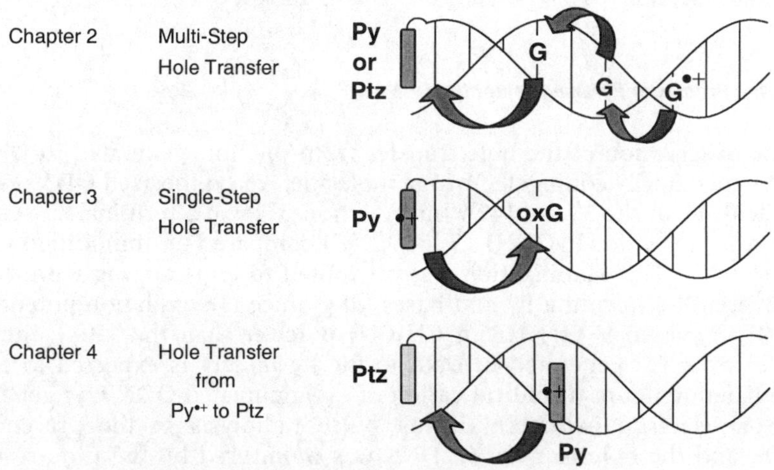

Scheme 2 Hole transfer processes described in this study

ing 7,8-dihydro-8-oxo-deoxyguanine (oxG) were used to investigate the single-step hole transfer from $Py^{\cdot+}$ to oxG [45]. In chapter 3, ODNs doubly conjugated with Py and Ptz were synthesized. Ptz was covalently bonded at the 5′-end, while Py was linked to the ribose 2′-site by the amide bond [47]. De-

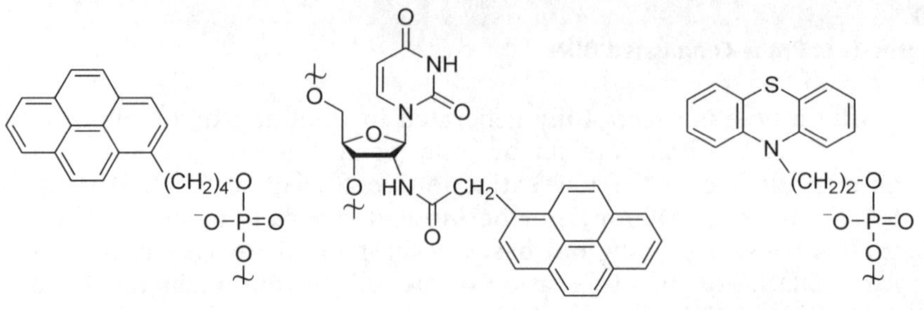

Fig. 1 Chemical structures of pyrene conjugated at the 5′-end (5′-Py) and the 2′ sugar position of uridine (PyU), and phenothiazine conjugated at the 5′-end of ODN (5′-Ptz)

pendences of the sequence and distance between Py and Ptz on the hole transfer from $Py^{\cdot +}$ to Ptz were investigated. The chemical structures of ODN conjugated with Py and Ptz and sequence of ODN used in this study are shown in Fig. 1.

2 Direct Observation of Hole Transfer in DNA

2.1 Multi-Step Processes

2.1.1 *Multi-Step Process in Pyrene-Conjugated ODN*

For the observation of the hole transfer from the hole generated in DNA to the site-specifically conjugated probe molecule, we synthesized ODNs conjugated with Py at the 5′-end [42]. Introduction of Py at a terminal site caused an increase in T_m for PyODN1 (T_m=39.1 °C) compared to unmodified ODN1 (T_m=34.2 °C). This stabilization was attributed to end-capping with π-electron overlap between the Py and bases [48]. Since the oxidation potential of Py [49] (E_{ox}=1.40 V vs NHE in CH_3CN) is lower than that of G, the hole transfer from $G^{\cdot +}$ generated in DNA to the Py moiety is expected to occur. The efficient amount of radical cation of Py-conjugated ODN was generated from one-electron oxidation during pulse radiolysis in the presence of $K_2S_2O_8$, and the hole transfer in DNA was monitored by the time-resolved transient absorption of $Py^{\cdot +}$ (Scheme 3) [45].

Distance dependence of the hole transfer process from the G-region (5′-GTGTGTG-3′) to the Py moiety was studied via pulse radiolysis of 5′-Py-conjugated ODNs with a different number of intervening A-T base pairs between the G-region and Py moiety (PyODN*n* (n=1~5)) (Scheme 3). Transient absorption with a maximum peak at 470 nm assigned to $Py^{\cdot +}$ was observed after the electron pulse during the pulse radiolysis (Fig. 2). This initial formation of $Py^{\cdot+}$ depended linearly on the concentration of PyODNn in the range of 0.05~0.2 mM, demonstrating that the formation of $Py^{\cdot+}$ in the time scale of 2 μs resulted from the direct diffusional collision between $SO_4^{\cdot-}$ and

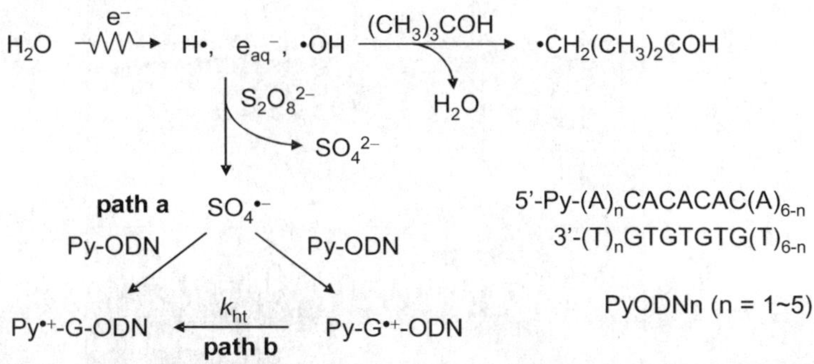

Scheme 3 Mechanistic scheme for generation of oxidizing reagent $SO_4^{\cdot-}$ during pulse radiolysis, hole generation and transfer in Py-conjugated ODN. Sequences of 5′-Py-conjugated ODN PyODN*n* (*n*=1–5)

the Py moiety (Scheme 3, path **a**). This gave the bimolecular reaction rate constant of the reaction of $SO_4^{\cdot-}$ with the Py moiety and ODN-region, $k_{Py\text{-}ODN}=4\times10^9\ M^{-1}\ s^{-1}$. In contrast, the rate constant of the reaction of $SO_4^{\cdot-}$ with unmodified ODN was determined to be $k_{ODN}=3\times10^9\ M^{-1}\ s^{-1}$. The simi-

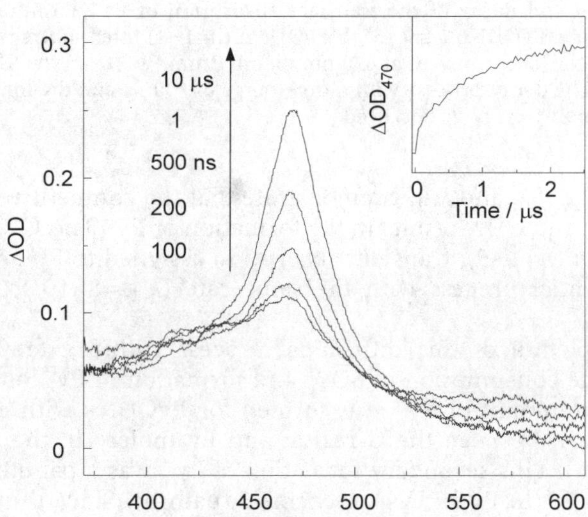

Fig. 2 Transient absorption spectra of PyODN1 observed several times after the electron pulse. The inset shows the time profile of the transient absorption peak of $Py^{\cdot+}$ at 470 nm, which corresponds to the direct oxidation of Py with $SO_4^{\cdot-}$ (Scheme 3, path **a**)

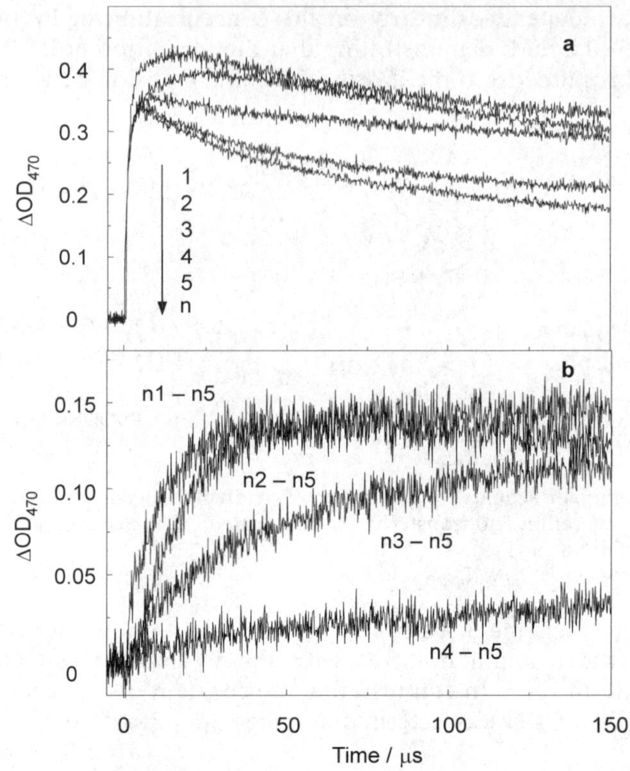

Fig. 3 Formation and decay of the transient absorption of $Py^{\cdot+}$ monitored at 470 nm. Time profiles for **a** PyODNn (n=1~5), **b** PyODNn (n=1~4) after subtraction of the time profile of PyODN5. The transient at 470 nm mainly from $Py^{\cdot+}$ from PyODNn decayed by two processes. The decay process with a time range of 100 μs and the longer component with a decay time >1 ms were observed

lar values of $k_{Py\text{-}ODN}$ and k_{ODN} demonstrate that the competitive oxidation of the Py moiety and ODN results in the formation of $Py^{\cdot+}$ and $G^{\cdot+}$, respectively. In all PyODNn (n=1~5), transient absorption assigned to $Py^{\cdot+}$ occurred by a single-exponential process with the same rate ($k_{obs}=8\times10^5$ s^{-1}) in a time range of 2 μs.

After completion of this diffusional process within a few μs, in other words after the consumption of $SO_4^{\cdot-}$ and formation of $Py^{\cdot+}$ or $G^{\cdot+}$, interestingly, a larger amount of $Py^{\cdot+}$ was formed for PyODNn with a shorter distance separation between the G-region and Py moiety in the time scale of 100 μs (Fig. 3a). This secondary formation of $Py^{\cdot+}$ was attributed to the hole transfer from $G^{\cdot+}$ to Py in DNA (Scheme 3, path **b**). Since the hole transfer from the G-region to the Py moiety is considered to be much slower for PyODN5 compared to the PyODNn (n=1~4), in the case of PyODN5 $Py^{\cdot+}$ was formed only from the initial collision process; the time profile of $Py^{\cdot+}$ at 470 nm for PyODN5 was subtracted from those for PyODNn (n=1~4). This

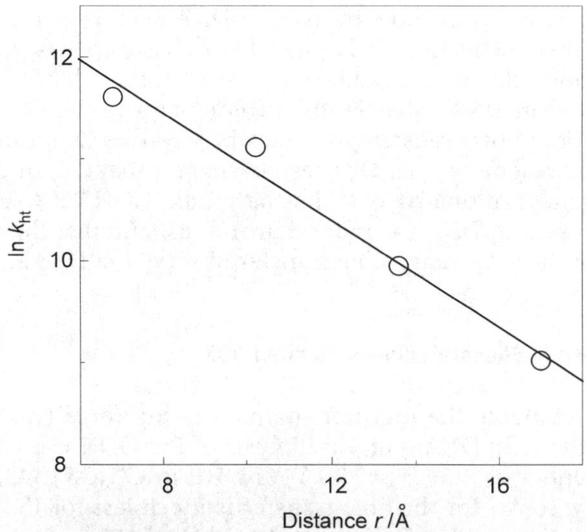

Fig. 4 Distance dependence of hole transfer in DNA. Shown is a semi-log plot of k_{ht} against the distance between Py and G-regions (r). Values of r were calculated assuming an average distance of 3.4 Å between base pairs

gave the time profiles of the formation of $Py^{\cdot+}$ from the hole transfer in DNA, and revealed the hole transfer in DNA over a period of 100 μs (Scheme 3, path **b**) (Fig. 3b). All calculated spectra were fitted by a single exponential function, and this gave the rate constant of the hole transfer (k_{ht}) of 11×10^4, 7×10^4, 2×10^4, and 0.8×10^4 s^{-1} for PyODNn (n=1, 2, 3, and 4, respectively). The hole transfer rate slightly decreased as the distance between the Py moiety and G-region (r) increased.

The rate constant for a single-step charge transfer process (k_{ct}) usually follows an exponential dependence on the donor-acceptor distance $r(k_{ct}=Ae^{-\beta r})$. When the natural logarithms of the hole transfer rate lnk_{ht} were plotted against r, a linear relationship was observed (Fig. 4). >From the slope of the plot, a value of 0.3 $Å^{-1}$ was obtained for the distance dependence of the hole transfer rate constant. The value of 0.3 $Å^{-1}$ is significantly smaller compared to the reported β values for the photo-induced single-step electron transfer process (β values between 0.6 to 1.4 have been reported for the single-step electron transfer in DNA [12, 13, 50]). The smaller value of 0.3 $Å^{-1}$ can be explained in that k_{ht} in the present study includes the multi-step hole transfer processes between Gs in the G-region. Likewise, the weak distance dependency has been reported for hole transfer with a multi-step process by Schuster et al (0.02 $Å^{-1}$) [4]. Since similar results were also observed when the concentration of DNA was decreased by half, the observed results are definitely accounted for by intramolecular hole transfer processes in DNA.

Deprotonation of $G^{\cdot+}$, which is considered to be fast ($>10^6$ s^{-1}), was suggested as the cause of the termination of the hole transfer in DNA. In contrast, the occurrence of the hole transfer in DNA over a period of 100 μs was demonstrated here. Therefore it is plausible to assume that proton transfer to hydrogen bonding complementary cytosine, rather than to surrounding water, is favored in DNA. That is, the proton stays in the $G^{\cdot+}$ and cytosine base pair, and the hole transfer processes between Gs are coupled with the proton transfer reactions [51]. Our results suggest that $G^{\cdot+}$, or a deprotonated G radical and protonated cytosine base-pair $G(-H^+)^{\cdot}$:$C(+H^+)^+$ [35, 52–56], can be present in DNA for more than 100 μs, and that this long lifetime is the origin of the long-range hole transfer observed in DNA.

2.2 Multi-Step Process in Phenothiazine-Conjugated ODN

In order to investigate the thermodynamic driving force (ΔG) dependence of the hole transfer in DNA, pulse radiolysis of Ptz-ODN was examined. The oxidation potential of Ptz (E_{ox}=0.76 V vs NHE in CH_3CN) [43, 44] is lower than that of Py, so ΔG for the hole transfer is much less for the hole transfer from $G^{\cdot+}$ to Ptz (ca. −0.1 eV) compared to that of $G^{\cdot+}$ to Ptz (ca. −0.7 eV). Similarly to $Py^{\cdot+}$, $Ptz^{\cdot+}$ has a distinct absorption peak with a large molar extinction coefficient as described above. 5′-Ptz-conjugated ODNs (PtzODNn) were synthesized by the automated solid-phase phosphoramidite method as reported [43, 44]. Introduction of Ptz at a terminal site caused an increase in T_m for PtzODNn (n=1, 3, 5) (T_m=36.7, 36.4, 35.7 °C) as compared to unmodified ODNn (T_m=35.5, 35.5, 35.4 °C, respectively). Since similar UV-spectra were observed for all PtzODNn (n=1, 3, 5), it was suggested that all PtzODNn take an end-capping structure with π-electron overlap between the Ptz and bases [48].

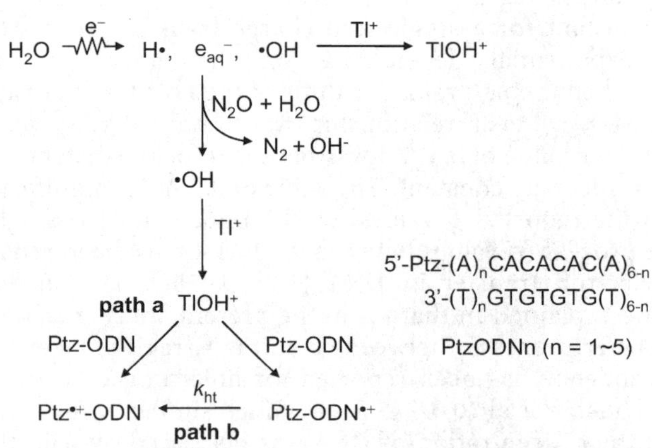

Scheme 4 Mechanistic scheme for generation of oxidizing reagent $TlOH^+$ during pulse radiolysis, hole generation and transfer in Ptz-conjugated ODN. Sequences of 5′-Ptz-conjugated ODN PtzODNn (n=1–5)

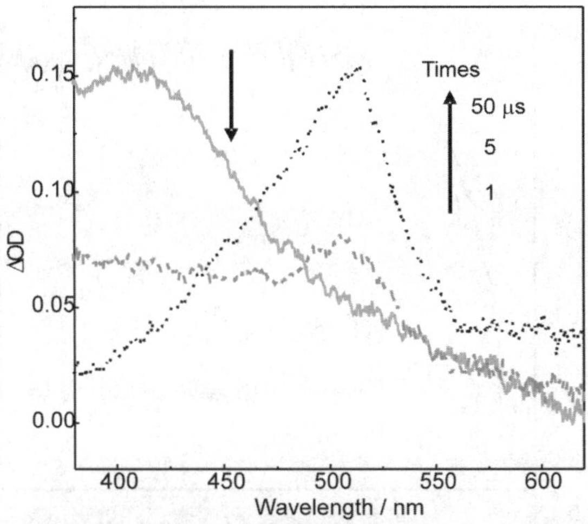

Fig. 5 Transient absorption spectra of PtzODN1 obtained at 1 μs, 5 μs, and 50 μs during pulse radiolysis of N_2O-saturated aqueous solution containing 0.2 mM PtzODN1 (strand conc.), 20 mM pH 7.0 Na phosphate buffer, and 1 mM Tl_2SO_4

Since one-electron oxidation of the Ptz moiety occurs with simple addition of $K_2S_2O_8$, Tl_2SO_4 was used as an oxidant source instead of $K_2S_2O_8$ in this study [40]. The oxidizing reagent $TIOH^+$ was generated during pulse radiolysis of an N_2O-saturated aqueous solution of Ptz-ODN. The distance dependence of the hole transfer was investigated systematically by changing the number of A-T base pairs separating the Ptz and G-region, as has been done for Py-conjugated ODNs (Scheme 4).

Transient absorption with a peak at 410 nm, assigned to $TIOH^+$, was observed after the electron pulse during the pulse radiolysis (Fig. 5). Together with the decay of $TIOH^+$, transient absorption with a maximum peak at 520 nm assigned to $Ptz^{\cdot+}$ was observed. In the case of PtzODN1, the observed formation rate of $Ptz^{\cdot+}$ and decay rate of $TIOH^+$ were almost the same (Fig. 6a). Therefore, the hole transfer from $G^{\cdot+}$ to Ptz in DNA is faster than the diffusional process for PtzODN1 ($k_{ht}>k_{obs}=1.4\times10^5$ s^{-1}).

After completion of the diffusional process within 20 μs – that is after the consumption of $TIOH^+$ and formation of $Ptz^{\cdot+}$ (Scheme 4, path **a**) or $G^{\cdot+}$ – secondary formation of $Ptz^{\cdot+}$ was observed for PtzODN3 over a time scale of 100 μs (Fig. 6b). This secondary formation of $Ptz^{\cdot+}$ was attributed to the hole transfer from $G^{\cdot+}$ to Ptz in DNA (Scheme 4, path b). The time profile of PtzODN3 was fitted by double exponentials. The faster component was assigned to the direct oxidation of Ptz by path **a**, while the slower component was assigned to the hole transfer process by path **b**. This gave the rate constant of the hole transfer $k_{ht}=2\times10^4$ s^{-1} for PtzODN3, which is similar to the hole transfer rate from a hole generated in DNA to Py for PyODN3 as mentioned above. No secondary formation was observed for PtzODN5, which

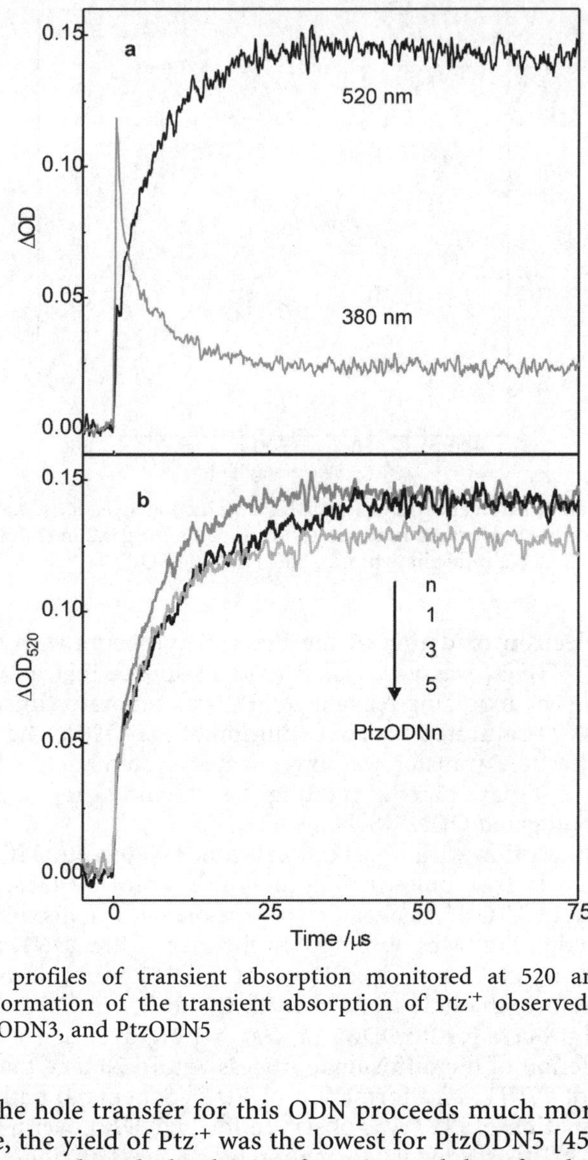

Fig. 6 **a** Time profiles of transient absorption monitored at 520 and 380 nm for PtzODN1 **b** Formation of the transient absorption of $Ptz^{\cdot+}$ observed at 520 nm for PtzODN1, PtzODN3, and PtzODN5

shows that the hole transfer for this ODN proceeds much more slowly. As a consequence, the yield of $Ptz^{\cdot+}$ was the lowest for PtzODN5 [45].

Therefore even though the driving force is much less for the hole transfer from $G^{\cdot+}$ to Ptz compared to that from $G^{\cdot+}$ to Py, similar kinetics were obtained. This is consistent with the assumption that the hole transfer from $G^{\cdot+}$ to Py or Ptz includes a multi-step process between Gs in the G-region: that is, the hole transfer from $G^{\cdot+}$ to Py or Ptz is not a rate determining step for the observed hole transfer.

2.3
Single-Step Process

In order to observe the single-step hole transfer from $Py^{\cdot +}$ to inside the DNA, pulse radiolysis of Py-conjugated ODNs with one oxG substituted for G as a hole trap was performed (Scheme 5; Fig. 7). In this case, the oxidation

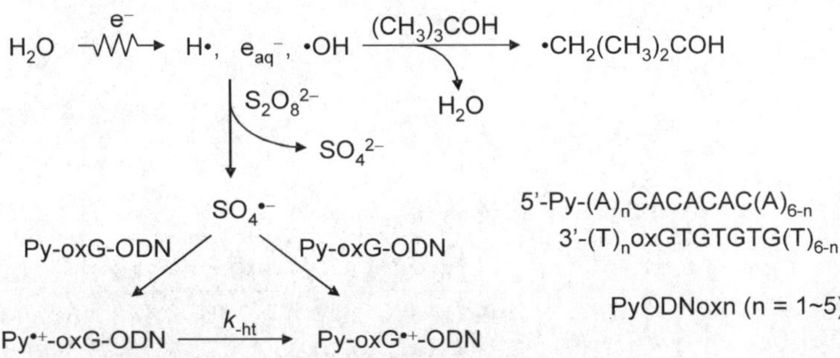

Scheme 5 Mechanistic scheme for generation of oxidizing reagent $SO_4^{\cdot -}$ during pulse radiolysis, hole generation and transfer in Py-conjugated ODN containing oxG. Sequences of 5′-Py-conjugated ODN containing oxG PyODNox*n* (*n*=1–5)

potential of oxG (E_{ox}=1.09 V vs NHE in CH_3CN) [57] is lower than that of Py, so hole transfer from $Py^{\cdot +}$ generated by the diffusional collision reaction with $SO_4^{\cdot -}$ to oxG was expected to occur. When the nearest G to the Py moiety was replaced with oxG (PyODNox1), the formation of $Py^{\cdot +}$ was suppressed. No secondary formation of $Py^{\cdot +}$ was observed for PyODNox*n* (*n*=2~5), which again confirms the hole transfer from the G-region to Py for the PyODN*n* (*n*=1~4).

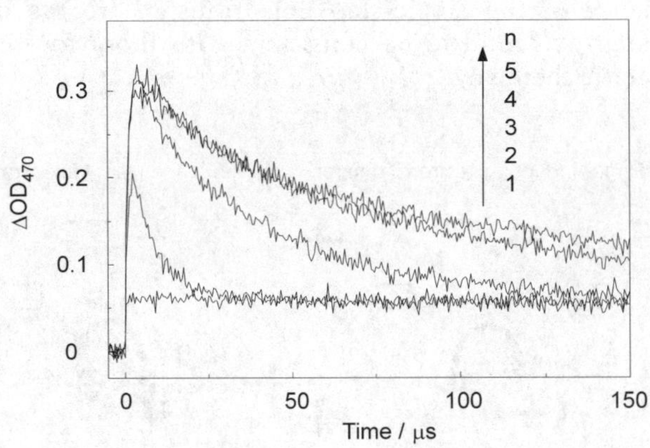

Fig. 7 Formation and decay of the transient absorption of $Py^{\cdot +}$ monitored at 470 nm for PyODNox*n* (*n*=1~5)

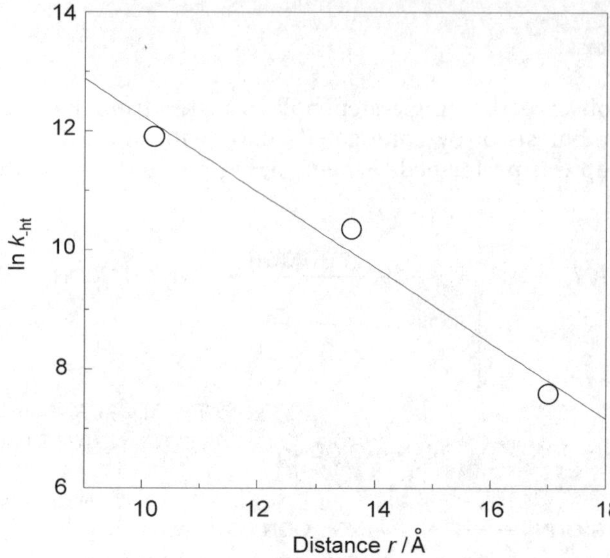

Fig. 8 Distance dependence of hole transfer in DNA. Shown is the semi-log plot of k_{-ht} against the distance between pyrene and oxG (r). Values of r were calculated assuming an average distance of 3.4 Å between base pairs

Since the decays of $Py^{\cdot+}$ of PyODNox$n$ (n=2~4) were the sum of the hole transfer from $Py^{\cdot+}$ to oxG and the decay of $Py^{\cdot+}$ with rate constants of k_{-ht} and k_{decay}, respectively, k_{-ht} was found to be 15×10^4, 3×10^4, and 0.2×10^4 s^{-1}, for PyODNox$n$ (n=2, 3, and 4, respectively). The rate constant of k_{-ht} decreased upon the increase of the distance between the Py and oxG. The linear plots of lnk_{-ht} vs r gave a slope, β=0.6 $Å^{-1}$ (Fig. 8), which is equal to the value of β for the photo-induced single-step electron transfer reported by Lewis et al and Zewail et al [12, 50]. It should be emphasized that the distance dependence of the single-step hole transfer process in DNA as a π-stacking system is found to be consistent with those for the single-step electron transfer (Scheme 6).

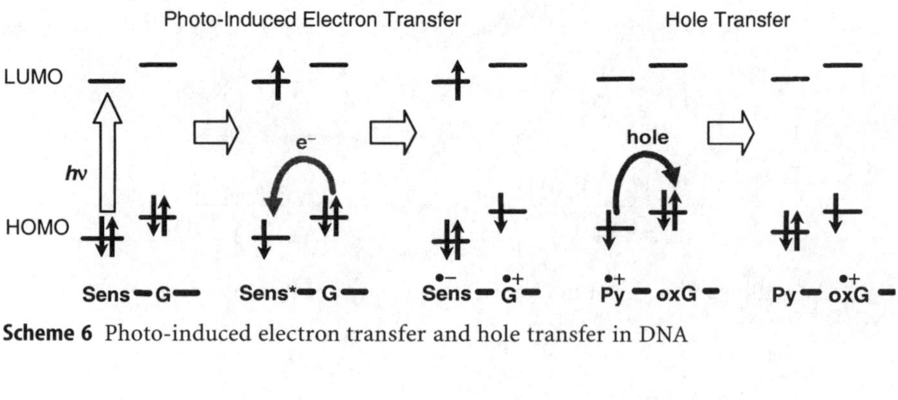

Scheme 6 Photo-induced electron transfer and hole transfer in DNA

2.4
Comparison with Photochemical Methods

Our results will now be compared with those based on transient absorption measurements during laser flash photolysis. Photochemical methods can also be used to generate a hole selectively in DNA. Using an intense 308 nm XeCl excimer laser, Shafirovich et al generated a radical cation of 2-amionpurine ($2AP^{\cdot+}$) in DNA [54, 56, 58, 59] during resonant two photon ionization. They measured the rate of the hole transfer from $2AP^{\cdot+}$ to oxG to be 4×10^4 and 3×10^3 s^{-1} in the case of two and four intervening A-T base pairs, respectively [54]. These values are similar to the rate of the hole transfer from $Py^{\cdot+}$ to oxG as mentioned above. Barton et al generated a hole in DNA by the oxidative flash quenching method using a dipyridophenazine complex of ruthenium as a tethered oxidant [60]. They monitored the formation of the 4-methylindole radical cation ($M^{\cdot+}$) which is positioned 17–37 Å away from the ruthenium complex. The rate of formation of the $M^{\cdot+}$ is $>10^7$ s^{-1}, which is much faster than Lewis' [17], Shafirovich's [54] or our results [45].

An advantage of the photochemical method is that a hole can be site-specifically generated in DNA within a short period of time. Since the quantum yield of the hole injection process is very low (usually lower than 0.001), the majority of photons absorbed by the chromophore are related to other photochemical phenomena. From this point of view, pulse radiolysis has an advantage, since radical cations can be generated in high yield in DNA. Absorption intensities of the radical cations observed in our studies are at least one order higher than those observed in the photochemical methods. However, disadvantages inherent to the pulse radiolysis are the requirement of high concentrations of DNA, and the limitation of the measurement to the time scale of >1 μs, because of the collisional process for the generation of a hole in DNA.

3
DNA as a Scaffold for Hole Transfer Between Two Organic Molecules: Hole Transfer from $Pyrene^{\cdot+}$ to Phenothiazine in DNA

Duplex DNA forming a one-dimensional π-stacked array can be used to construct highly-ordered structures in nanoscale devices [61–63]. Evidence of long-range hole transfer over a distance of 200 Å suggests that DNA may be used as nanowires for long range charge transportation [9]. However, DNA is inherently unstable upon oxidation [7, 8, 34], so DNA itself cannot be a good conductive material. However, there are organic molecules whose radical cations are stable in water. Therefore, for the construction of DNA wires, it may be better to use DNA as a scaffold for arranging such organic molecules.

In sections 2.1 and 2.2, Py and Ptz were demonstrated to be useful as hole carrying molecules for DNA. With this in mind, we now describe the pulse radiolysis of ODN doubly conjugated with Py and Ptz, carried out in order to investigate the possibility of modulation of the hole transfer rate and direction by means of introducing several organic molecules into DNA. Monitoring the transient absorption of $Py^{\cdot+}$ and $Ptz^{\cdot+}$, we measured the dependency of the hole transfer rate on the distance and sequence in DNA.

Planar Py and Ptz are expected to stack well among nucleobases, and so were used as the probe molecules to measure the hole transfer rate. The oxidation potentials of Py and Ptz were lower than those of the four nucleobases. We examined the sequence and distance dependences of the hole transfer rate by monitoring the decay and formation of the transient absorption of $Py^{\cdot+}$ and $Ptz^{\cdot+}$ during pulse radiolysis of ODN conjugated with Py and Ptz (PtzPy-(***n***) (***n***=1~5)).

The introduction of Py at the 2′ sugar position of uridine and Ptz at the 5′-end of ODN caused an increase in T_m. The T_m for PtzPy-**1** is 39.4 °C, which is 6.9 °C higher than that of unmodified ODN (32.5 °C). Similarly, the introduction of Ptz or Py into ODN showed increases in T_m for Ptz-**1** (1.2 °C) and for Py-**1** (6.6 °C) compared with unmodified ODN, suggesting that Py intercalated into ODN duplex at the 3′-side, and the 5′-linked-Ptz associated with the 5′-terminus by end-capping [5]. The structures of ODNs conjugated with Py and Ptz were examined by circular dichroism (CD) spectral measurements. The CD spectra of ODNs in 20 mM phosphate buffer were characteristic of the B-form.

Oxidizing reagent $TlOH^+$ was generated by pulse radiolysis and direct oxidation of ODN by $TlOH^+$ was completed within 20 μs [46] (Scheme 7, path **a**).

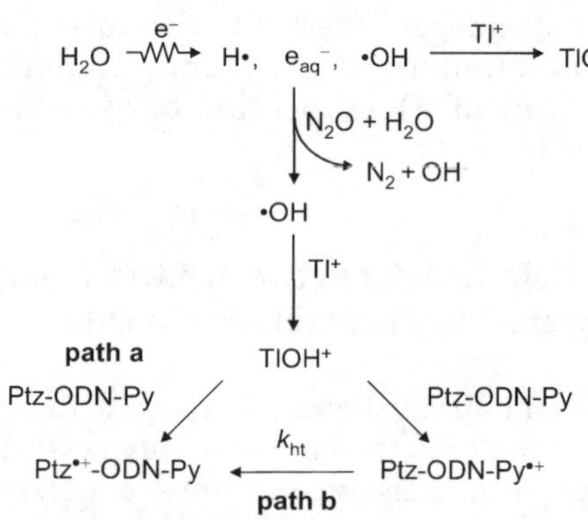

Scheme 7 Mechanistic scheme for generation of oxidizing reagent $TlOH^+$ during pulse radiolysis, hole generation and transfer in Ptz- and Py-conjugated ODN Ptz-ODN-Py

The pulse radiolysis of ODN conjugated with Py and Ptz showed a transient absorption spectrum with a peak at 470 or 520 nm, assigned to $Py^{\cdot+}$ or $Ptz^{\cdot+}$ respectively (Fig. 9). The formation rates of $Py^{\cdot+}$ and $Ptz^{\cdot+}$ obtained for Py-**1** and Ptz-**1** were almost identical to that of the decay of $TlOH^+$ and no secondary formation was observed. These results indicate that a hole migrates from

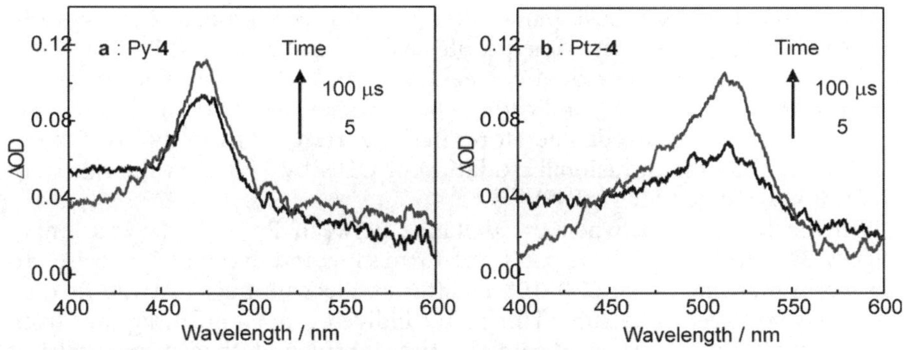

Fig. 9 Transient absorption spectra observed at 5 and 100 μs after the electron pulse during pulse radiolyses of **a** Py-4 and **b** Ptz-4 in N_2O-saturated aqueous solution containing 0.2 mM ODN (strand conc.), 20 mM Na phosphate buffer (pH 7.0), and 2 mM Tl_2SO_4

the G-region to the Py or Ptz moiety within the diffusional process. Considering lower oxidation potentials of Py and Ptz compared to that of G, a hole generated in ODN migrates from G to Py or Ptz, and it is expected that this process occurred within several μs according to our previous reports [45, 46, 64]. Therefore, contribution of the hole transfer from the G-region was almost ignored in the examined time region (>10 μs). As a result, a hole is trapped on the Py- and Ptz-sites after the consumption of oxidant $TlOH^+$.

Transient absorption spectra changes and time profiles observed during pulse radiolysis of PtzPy-**n** are shown in Fig. 10. When Ptz was separated

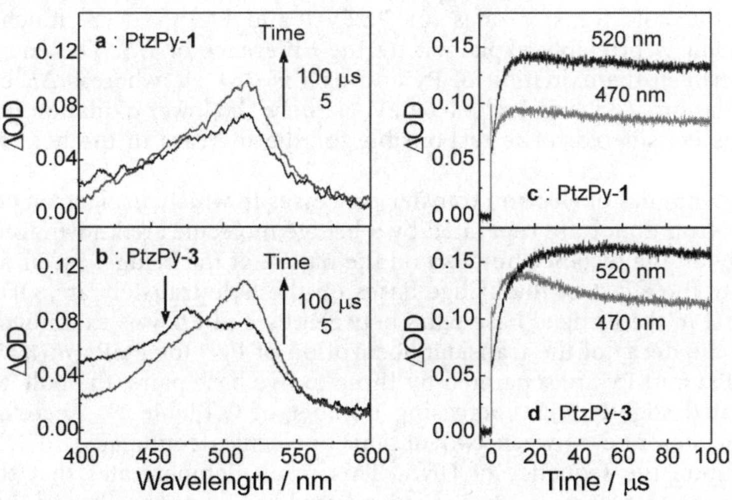

Fig. 10 Transient absorption spectra observed at 5 and 100 μs after the electron pulse during pulse radiolyses of **a** PtzPy-1 and **b** PtzPy-3. Also shown are time profiles of the transient absorption of $Py^{\cdot+}$ and $Ptz^{\cdot+}$ observed at 470 and 520 nm for **c** PtzPy-1 and **d** PtzPy-3, respectively

from Py by three A-T base pairs (PtzPy-**1**), no formation of Py$^{\cdot+}$ was observed (Fig. 10a). From the time profiles of Ptz$^{\cdot+}$ for Ptz-**1** and Py$^{\cdot+}$ for Py-**1**, the rate constants of the oxidation reaction with TIOH^{+} were determined to be approximately 10^5 s^{-1}, indicating that competitive oxidation of the Ptz and Py-moieties occurred. Therefore, the hole transfer from Py$^{\cdot+}$ to Ptz occurs faster than the diffusional oxidation of ODN by TIOH^{+} at the rate constant of $k_{obs}=10^5$ s^{-1} for PtzPy-**1**.

On the other hand, when the distance between Py and Ptz was longer (PtzPy-**3**) with five A-T base pairs, the formation and decay of Py$^{\cdot+}$ were observed in the time range of 0–100 μs after an electron pulse. At 100 μs only Ptz$^{\cdot+}$ was observed (Fig. 10b). This result indicates that a hole migrates from Py$^{\cdot+}$ to Ptz within 100 μs. >From the time profiles of absorption peaks at 470 nm for Py$^{\cdot+}$ and 520 nm for Ptz$^{\cdot+}$, secondary formation of Ptz$^{\cdot+}$ was also observed concomitant with the decay of Py$^{\cdot+}$ for PtzPy-**3**, while no secondary formation of Ptz$^{\cdot+}$ was observed for PtzPy-**1** due to the rapid hole transfer (Fig. 10c,d). The rate constant of the hole transfer from Py$^{\cdot+}$ to Ptz was determined to be 2.0×10^4 s^{-1} for PtzPy-**3** from the decay of Py$^{\cdot+}$.

These results demonstrate that the hole transfer rate decreases with the increase of the distance between Py and Ptz (Scheme 7, path **b**). When Ptz and Py were separately conjugated to different ODNs, both Py$^{\cdot+}$ and Ptz$^{\cdot+}$ were observed at 100 μs after the electron pulse during pulse radiolysis of the mixture of 0.2 mM each of Py- and Ptz-conjugated ODNs: in other words no interstrand hole transfer occurred. Therefore, the observed results are accounted for by intramolecular processes.

In section 2.2, we reported that the rate constants of the hole transfer from Py$^{\cdot+}$ to oxG were 3×10^4 s^{-1} for Py-TTT-oxG and smaller than 1×10^3 s^{-1} for Py-TTTTT-oxG [45]. Compared with the rate constants for Py-oxG-ODNs, the hole transfer rates for PtzPy-**1** and PtzPy-**3** were much faster. This result is probably explained by the difference of ΔG. The energy gap (ΔE) between the oxidation of Py and oxG is 0.31 eV, whereas ΔE between the oxidation of Py and Ptz is 0.64 eV. Namely, the lower oxidation potential of Ptz is considered to be responsible for the increase in the hole transfer rate.

The dynamics of electron transfer processes in which an electron acceptor and electron donor are separated by a bridge molecule are known to depend not only on the distance but also on the nature of the bridge. So, in order to elucidate the effect of the bridge bases on the hole transfer rate, ODN containing G in the bridged base pairs between Py and Ptz was examined. Monitoring the decay of the transient absorption of Py$^{\cdot+}$ for PtzPy-$\boldsymbol{n}$ ($\boldsymbol{n}$=**3**~**5**) in which Ptz and Py are separated by three to five base pairs, the hole transfer accelerated slightly with increasing number of G (Table 1). Therefore, the hole transfer rate between two organic molecules conjugated to DNA depends upon the sequence of DNA. This result demonstrates that the hole transfer process from Py$^{\cdot+}$ to Ptz is mediated by G between Py and Ptz, indicating increases of the electronic coupling for the superexchange interaction. Similar results have been reported by Saito et al. using strand cleavage anal-

Table 1 Sequences of ODN conjugated with Ptz and Py (PtzPy-*n*), and the rate Constant of the hole transfer from Py to Ptz for PtzPy-*n*

ODN	Sequence	k_{ht}[a] (10^4 s^{-1})
PtzPy-1	$^{5'}$**Ptz**-AAA- A-CGCGATA TTT-**PyU**-GCGCTAT$^{5'}$	>10
PtzPy-2	$^{5'}$**Ptz**-ACA- A-CGCGATA TGT-**PyU**-GCGCTAT$^{5'}$	>10
PtzPy-3	$^{5'}$**Ptz**-AAAAA- A-CGCGA TTTTT-**PyU**-GCGCT$^{5'}$	2.0
PtzPy-4	$^{5'}$**Ptz**-AACAA- A-CGCGA TTGTT-**PyU**-GCGCT$^{5'}$	2.6
PtzPy-5	$^{5'}$**Ptz**-ACACA- A-CGCGA TGTGT-**PyU**-GCGCT$^{5'}$	3.3

[a] Obtained from the decay of the transient absorption of $Py^{\cdot+}$ at 470 nm

ysis [65]. However, the contribution of G as a stepping stone for the hole transfer may also be considered.

So, to summarize, the hole transfer rate between Py and Ptz was determined using the transient absorption measurements of $Py^{\cdot+}$ and $Ptz^{\cdot+}$, and we have shown that the hole transfer rate from $Py^{\cdot+}$ to Ptz depends on the distance and sequence between Ptz and Py. In other words, the hole transfer rate in DNA is modulated by designing the sequence of DNA and conjugated molecules. It is suggested that DNA may be utilized as a molecular wire by introducing several organic molecules at the appropriate site in DNA.

4 Conclusion

Investigations of the kinetics of hole transfer in DNA by means of pulse radiolysis of synthetic ODNs have provided details about the hole transfer process, especially over 1 μs, including the multi-step hole transfer process. Based on the investigation of the kinetics of hole transfer in DNA, development of the DNA nanoelectronic devices is now expected. An active application of the hole transfer process is also desirable from a therapeutical point of view, since hole transfer may play a role in improvement of quantum yield and selectivity of DNA scission during photodynamic therapy. The kinetics of the hole transfer process is now being revealed, although there is still much research to be performed in this area. The kinetics of adenine hopping is another area of interest that should be explored in the future.

Acknowledgement We are deeply indebted to Mr. Tadao Takada and Mrs. Sachiko Tojo who have contributed to the experiment. We also thank the members of the Radiation Laboratory of SANKEN, Osaka University, for running the linear accelerator. This work has been partly supported by a Grant-in-Aid for Scientific Research from the Ministry of Education, Science, Sport and Culture of Japan.

References

1. Seidel CAM, Schulz A, Sauer MHM (1996) J Phys Chem 100:5541
2. Meggers E, Michel-Beyerle ME, Giese B (1998) J Am Chem Soc 120:12950
3. Giese B (2000) Acc Chem Res 33:631
4. Henderson PT, Jones D, Hampikian G, Kan YZ, Schuster GB (1999) Proc Natl Acad Sci USA 96:8353
5. Schuster GB (2000) Acc Chem Res 33:253
6. Kelley SO, Barton JK (1998) Chem Biol 5:413
7. Burrows CJ, Muller JG (1998) Chem Rev 98:1109
8. Armitage B (1998) Chem Rev 98:1171
9. Porath D, Bezryadin A, de Vries S, Dekker C (2000) Nature 403:635
10. Murphy CJ, Arkin MR, Jenkins Y, Ghatlia ND, Bossmann SH, Turro NJ, Barton JK (1993) Science 262:1025
11. Meade TJ, Kayyem JF (1995) Angew Chem Int Ed Engl 34:352
12. Lewis FD, Wu TF, Zhang YF, Letsinger RL, Greenfield SR, Wasielewski MR (1997) Science 277:673
13. Fukui K, Tanaka K (1998) Angew Chem Int Ed Engl 37:158
14. Kelley SO, Barton JK (1999) Science 283:375
15. Saito I, Takayama M, Sugiyama H, Nakatani K, Tsuchida A, Yamamoto M (1995) J Am Chem Soc 117:6406
16. Sugiyama H, Saito I (1996) J Am Chem Soc 118:7063
17. Lewis FD, Liu XY, Liu JQ, Miller SE, Hayes RT, Wasielewski MR (2000) Nature 406:51
18. Calcaterra LT, Closs GL, Miller JR (1983) J Am Chem Soc 105:670
19. Miller JR, Calcaterra LT, Closs GL (1984) J Am Chem Soc 106:3047
20. Miller JR, Penfield K, Johnson M, Closs G, Green N (1998) Adv Chem Ser 254:161
21. Yamamoto Y (1992) Trends Org Chem 3:93
22. Tamai T, Mizuno K, Hashida I, Otsuji Y, Ishida A, Takamuku S (1994) Chem Lett 149
23. Kimura N, Takamuku S (1994) J Am Chem Soc 116:4087
24. Tsuchida A, Ikawa T, Yamamoto M, Ishida A, Takamuku S (1995) J Phys Chem 99:14793
25. Kimura N, Takamuku S (1995) J Am Chem Soc 117:8023
26. Tojo S, Morishima K, Ishida A, Majima T, Takamuku S (1995) Bull Chem Soc Jpn 68:958
27. Ishida A, Fukui M, Ogawa H, Tojo S, Majima T, Takamuku S (1995) J Phys Chem 99:10808
28. Majima T, Tojo S, Ishida A, Takamuku S (1996) J Org Chem 61:7793
29. Majima T, Tojo S, Ishida A, Takamuku S (1996) J Phys Chem 100:13615
30. Majima T, Fukui M, Ishida A, Takamuku S (1996) J Phys Chem 100:8913
31. Fujita M, Ishida A, Majima T, Takamuku S (1996) J Phys Chem 100:5382
32. Kojima M, Kakehi A, Ishida A, Takamuku S (1996) J Am Chem Soc 118:2612
33. Majima T, Tojo S, Takamuku S (1997) J Phys Chem B 101:1048
34. Steenken S (1989) Chemical Reviews 89:503
35. Candeias LP, Steenken S (1989) J Am Chem Soc 111:1094
36. Candeias LP, Wolf P, Oneill P, Steenken S (1992) J Phys Chem 96:10302
37. Steenken S, Telo JP, Novais HM, Candeias LP (1992) J Am Chem Soc 114:4701
38. Steenken S, Jovanovic SV (1997) J Am Chem Soc 119:617
39. Appleby A, Schwarz HA (1969) J Phys Chem 73:1937
40. Schwarz HA, Dodson RW (1984) J Phys Chem 88:3643
41. Alkaitis SA, Beck G, Gratzel M (1975) J Am Chem Soc 97:5723
42. Mann JS, Shibata Y, Meehan T (1992) Bioconjugate Chem 3:554
43. Tierney MT, Sykora M, Khan SI, Grinstaff MW (2000) J Phys Chem 104:7574
44. Tierney MT, Grinstaff MW (2000) J Org Chem 65:5355
45. Kawai K, Takada T, Tojo S, Ichinose N, Majima T (2001) J Am Chem Soc 123:12688
46. Kawai K, Takada T, Tojo S, Majima T (2002) Tetrahedron Lett 43:89
47. Yamana K, Asazuma K, Nakano H (1999) Nucleic Acids Res Symp Ser 42:113

48. Guckian KM, Schweitzer BA, Ren RXF, Sheils CJ, Paris PL, Tahmassebi DC, Kool ET (1996) J Am Chem Soc 118:8182
49. Murov SL, Carmichael I, Hug GL (1993) Handbook of Photochemistry, 2nd edn. Marcel Dekker, New York
50. Wan CZ, Fiebig T, Schiemann O, Barton JK, Zewail AH (2000) Proc Natl Acad Sci USA 97:14052
51. Kawai K, Majima T (2002) J Photo Photo C 3:53
52. Hutter M, Clark T (1996) J Am Chem Soc 118:7574
53. Nir E, Kleinermanns K, de Vries MS (2000) Nature 408:949
54. Shafirovich V, Cadet J, Gasparutto D, Dourandin A, Huang WD, Geacintov NE (2001) J Phys Chem 105:586
55. Giese B, Wessely S (2001) Chem Comm 2108
56. Shafirovich V, Dourandin A, Geacintov NE (2001) J Phys Chem 105:8431
57. Steenken S, Jovanovic SV, Bietti M, Bernhard K (2000) J Am Chem Soc 122:2373
58. Shafirovich V, Dourandin A, Huang W, Luneva NP, Geacintov NE (1999) Photochem Photobiol 69:62s
59. Shafirovich V, Dourandin A, Huang WD, Luneva NP, Geacintov NE (2000) Phys Chem Chem Phys 2:4399
60. Pascaly M, Yoo J, Barton JK (2002) J Am Chem Soc 124:9083
61. Winfree E, Liu FR, Wenzler LA, Seeman NC (1998) Nature 394:539
62. Braun E, Eichen Y, Sivan U, Ben-Yoseph G (1998) Nature 391:775
63. Storhoff JJ, Mirkin CA (1999) Chem Rev 99:1849
64. Kawai K, Takada T, Tojo S, Majima T (2002) Tetrahedron Lett 43:8083
65. Nakatani K, Dohno C, Saito I (2000) J Am Chem Soc 122:5893

Top Curr Chem (2004) 236:139–161
DOI 10.1007/b94414

The Mechanism of Long-Distance Radical Cation Transport in Duplex DNA: Ion-Gated Hopping of Polaron-Like Distortions

Gary B. Schuster · Uzi Landman

School of Chemistry & Biochemistry and the School of Physics, Georgia Institute of Technology, Atlanta, GA 30332, USA
E-mail: gary.schuster@cos.gatech.edu

Abstract The irradiation of an anthraquinone derivative that is covalently attached to duplex DNA injects a radical cation into the bases of the DNA. This radical cation can migrate hundreds of Ångstroms before it is trapped at GG steps by reaction with water. These damaged guanines result in DNA strand scissions when they are treated with piperidine. Investigation of several such DNA constructs reveals that the efficiency of radical cation migration is strongly dependent on the sequence of bases in the DNA. This observation led to the formulation of the phonon-assisted polaron hopping model for the mechanism of radical cation migration. In this model, DNA and its ionic and solvent environment are assumed to undergo motions on the timescale of the radical cation hopping. These motions lead to a distortion of the local environment around the radical cation that causes it to gain stability (the polaron). Thermal motions of the DNA and its environment (phonons) cause the radical cation to migrate adiabatically from one polaronic site to another.

Keywords Long-distance charge transport · DNA damage · Polaron hopping · Ion gated base sequence effects

1 Introduction

DNA is a chemical whose function in living cells is to store the instructions needed to maintain life. Errors introduced into those instructions generally have deleterious consequences, so there is great evolutionary pressure to prevent or correct them. At the molecular level, a reaction that changes the structure of DNA damages the instructions and introduces errors. Several reactions cause structural changes in DNA; among the most important is one-electron oxidation [1]. Oxidation of DNA can result from normal cellular metabolism, from exposure to ionizing radiation, or from interaction with light [2–5].

When DNA is oxidized, it loses an electron and a radical cation ("hole") is generated. Overwhelming evidence shows that these radical cations reside primarily on the aromatic bases that form the central core of duplex DNA. Radical cations in DNA are short-lived species that are consumed by reaction with H_2O to produce structurally modified (damaged) bases [6]. It is now widely accepted that the base initially oxidized is not necessarily the base that is eventually damaged by reaction of the radical cation with H_2O [7–9]. The radical cation migrates through the DNA duplex until it is eventually consumed by a reaction. Clearly, it is essential to understand the mechanism for long-distance migration of radical cations in DNA because that process controls the site of oxidative damage. Our findings on that topic are described in this chapter.

2 Anthraquinones as One-Electron Oxidants of DNA

Redox reactions follow well-established thermodynamic and kinetic principles. Generally, a one-electron oxidation reaction is spontaneous and rapid when its driving force ($-\Delta G_{et}$) is greater than about 5 kcal/mol (0.2 eV) and the electron donor and acceptor are at a near contact distance. Electronically excited states formed by optical excitation are often powerful oxidants. In this case, the Weller equation provides a convenient means to estimate ΔG_{et} based on the energy of the excited state (ΔE^*), the oxidation potential (E_{ox}) of the electron donor (a DNA base in the present case), the reduction potential of the electron acceptor (E_{red}), and certain electrostatic work terms [10]. Numerous organic and metallorganic compounds have been found whose excited state meets the energetic requirement for oxidation of DNA. We have focused our attention on anthraquinone derivatives.

Anthraquinones are nearly perfect sensitizers for the one-electron oxidation of DNA. They absorb light in the near-UV spectral region (350 nm) where DNA is essentially transparent. This permits excitation of the quinone without the simultaneous absorption of light by DNA, which would confuse chemical and mechanistic analyses. Absorption of a photon by an anthraquinone molecule initially generates a singlet excited state; however, intersystem crossing is rapid and a triplet state of the anthraquinone is normally formed within a few picoseconds of excitation, see Fig. 1 [11]. Application of the Weller equation indicates that both the singlet and the triplet excited states of anthraquinones are capable of the exothermic one-electron oxidation of any of the four DNA bases to form the anthraquinone radical anion ($AQ^{-\cdot}$) and a base radical cation ($B^{+\cdot}$).

Oxidation reactions that originate with the singlet excited state of the anthraquinone (AQ^{*1}) generate a contact radical ion pair in an overall singlet

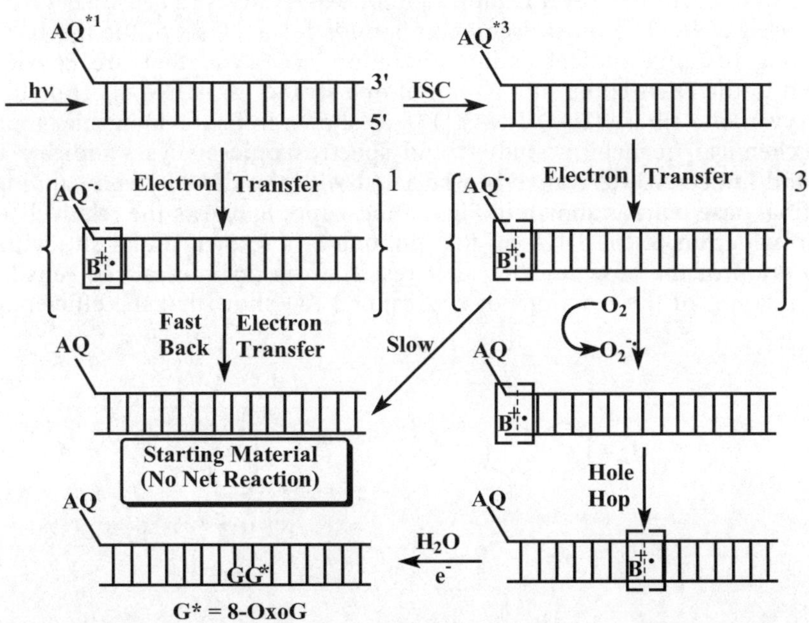

Fig. 1 Schematic mechanism for the long-distance oxidation of DNA. Irradiation of the anthraquinone (AQ) and intersystem crossing (ISC) forms the triplet excited state (AQ^{*3}), which is the species that accepts an electron from a DNA base (B) and leads to products. Electron transfer to the singlet excited state of the anthraquinone (AQ^{*1}) leads only to back electron transfer. The anthraquinone radical anion ($AQ^{-\cdot}$) formed in the electron transfer reaction is consumed by reaction with oxygen, which is reduced to superoxide. This process leaves a base radical cation ($B^{+\cdot}$, a "hole") in the DNA with no partner for annihilation, which provides time for it to hop through the DNA until it is trapped by water (usually at a GG step) to form a product, 7,8-dihydro-8-oxoguanine (8-OxoG)

spin state that can undergo rapid back electron transfer to regenerate the starting materials. Our findings indicate that this unproductive charge annihilation route dominates reactions that originate from AQ^{*1} [11].

On the other hand, oxidation of a DNA base by a triplet state of the anthraquinone (AQ^{*3}) generates a contact ion pair in an overall triplet state, and back electron transfer from this species to form ground states is prohibited by spin conservation rules. Consequently, the lifetime of the triplet radical ion pair is long enough to permit the bimolecular reaction of $AQ^{-\cdot}$ with O_2 to form superoxide ($O_2^{-\cdot}$) and regenerate the anthraquinone.

Therefore, the sequence of reactions illustrated in Fig. 1 catalytically (the anthraquinone is regenerated) "injects" a radical cation into a DNA oligonucleotide that does not simultaneously contain a radical anion. As a result, the lifetime of this radical cation is determined by its relatively slow bimolecular reaction with H_2O (or some other diffusible reagent such as $O_2^{-\cdot}$) and not by a rapid intramolecular charge annihilation reaction. This provides sufficient time for the long distance migration of the radical cation in DNA to occur.

We have examined several anthraquinone derivatives as sensitizers for oxidation of DNA. The most useful compounds for analysis of the mechanism for long distance radical cation migration are those that are covalently linked to the DNA either at a 5′-end of one strand (AQ-DNA) [12] or to the 3′-oxygen of a ribose (UAQ-DNA) [13], as shown in Fig. 2. Molecular modeling, chemical quenching studies, and spectroscopic analyses indicate that the end-linked AQ derivative is associated with the DNA by end-capping of the final base pair, as shown in Fig. 3. End-capping allows the relatively efficient oxidation of the DNA by the quinone at a known initial site without disruption of the base stacking that results from an intercalated sensitizer. Examination of the reactions of end-capped AQ show that the efficiency of

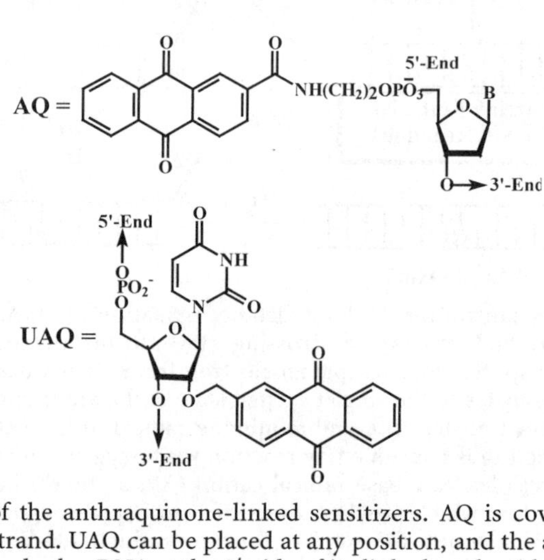

Fig. 2 Structures of the anthraquinone-linked sensitizers. AQ is covalently attached to the 5′-end of one strand. UAQ can be placed at any position, and the attached anthraquinone intercalates in duplex DNA at the 3′-side of its linked nucleotide

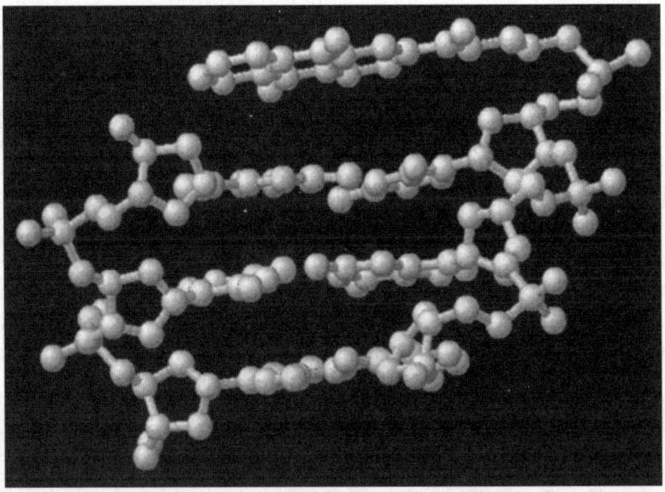

Fig. 3 Model of an end-capped anthraquinone that is covalently linked to a 5′-terminus of duplex DNA by the tether shown in Figure 2

charge injection depends on the sequence of bases near the AQ. Maximum efficiency is observed when there is no G/C base pair within the three base pairs closest to the AQ [14]. The sequence effect on charge injection efficiency is attributed to more rapid migration of the base radical cation away from the quinone radical anion when there is no nearby guanine, which acts as a shallow trap.

The anthraquinone group of the UAQ sensitizer is intercalated on the 3′-side of its linkage site [15]. Use of UAQ permits assessment of the directionality of long-range radical cation migration. Both AQ and UAQ enable the selective and efficient introduction of a radical cation in duplex DNA, whose lifetime is controlled by its relatively slow bimolecular reaction primarily with H_2O.

3
Interpretation of Radical Cation Reaction Patterns in Duplex DNA

Irradiation of an AQ-linked duplex DNA oligomer leads to selective reaction at certain base pair sequences. This reaction is detected as strand cleavage, after treatment of the irradiated sample with piperidine, by polyacrylamide gel electrophoresis (PAGE) on DNA oligomers that contain a ^{32}P radiolabel. This behavior is indicative of chemical reaction (damage) at a DNA bases rather than at a deoxyribose sugar, in which case strand cleavage generally does not require treatment with piperidine [5]. It is typically found by us and by others (using a variety of oxidants) that reaction of the radical cation usually occurs primarily at the 5′-G of GG [16, 17] steps (or at the central and 5′-G of GGG) and less frequently at the G of a 3′-AG-5′ sequence. This selectivity of reaction has been attributed to stabilization of the radical cation at GG or AG sites due to delocalization of the charge [18]. However, stabilization by GG steps does not generate a "deep trap" from which the radical cation cannot escape, since reaction is routinely observed at GG steps that are near to the AQ sensitizer (proximal) and at those further away (distal) so that the radical cation must pass through the proximal GG step to cause reaction at the distal site [12, 19].

Analysis of the relative efficiency of strand cleavage of duplex DNA provides useful information on the relative rates of charge transport, that permits analysis of the mechanism for radical cation migration. These experiments must be carried out under conditions of low conversion ("single hit") so that each DNA oligomer, on average, reacts once or not at all. Under these conditions, the competition between the rate of reaction of the radical cation with H_2O and its migration is revealed by the statistical pattern of the cleavage results. This is illustrated by considering two limiting examples.

In the first case, we presume that the rate of reaction of H_2O with the radical cation is much faster than the rate of its migration. In this case, reaction will be observed only at the GG step closest to the covalently linked AQ; the radical cation never reaches distal GG steps.

In the second limiting case, the rate of reaction with H_2O is presumed to be much slower than the rate of radical cation migration and independent of the specific base pair sequence surrounding the GG step. Under these circumstances, each GG step will be equally reactive, and just as much strand cleavage will be observed at the GG step farthest from the AQ as at the one closest to it.

In the intermediate circumstance where the rate of reaction with H_2O and the rate of radical cation migration are comparable, then the amount of reaction detected is somehow related to the distance from the AQ to the GG step.

Therefore, analysis of the efficiency and pattern of strand cleavage provides information on the relative rate of radical cation migration through different DNA sequences. This is powerful information for analysis of the charge migration mechanism.

4
The Base Sequence and Distance Dependence of Radical Cation Migration

The pattern and efficiencies of strand cleavage at GG steps in duplex DNA reflect the ability of a radical cation to migrate from its initial position through a sequence of base pairs. In an illustrative example, we consider the photochemistry of AQ-DNA(1), which is shown in Fig. 4. AQ-DNA(1) is a 20-mer that contains an AQ group linked to the 5′-end of one strand and has two GG steps in the complementary strand. The proximal GG step is eight base pairs, ca. 27 Å, from the 5′-end linked to the AQ, and the distal GG step is 13 base pairs (ca. 44 Å) away. The complementary strand is labeled with ^{32}P at its 5′-terminus (indicated by a * in Fig. 4).

Measurement of the melting temperature (T_m) and the circular dichroism (CD) spectrum of AQ-DNA(1) shows that it is a duplex at room temperature

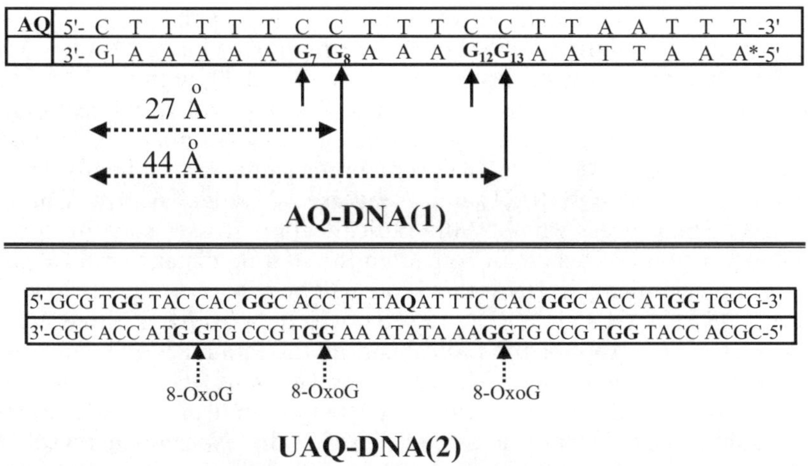

Fig. 4 Schematic representation of long-distance radical cation migration in DNA. In AQ-DNA(1), irradiation of the anthraquinone group linked at the 5′-terminus leads to reaction at GG steps that are 27 Å and 44 Å from the site of charge injection. The amount of reaction observed at each guanine is represented approximately by the length of the solid arrow. In UAQ-DNA(2), irradiation of the anthraquinone leads to reaction at each of the eight GG steps. However, replacement of a G by 7,8-dihydro-8-oxoguanine (8-OxoG) introduces a deep trap that inhibits reaction at guanines on the same side of the DNA as the trap

in 10 mM sodium phosphate buffer solution at pH 7, which are the standard conditions we use for the irradiation experiments. Irradiation of a 2.5 μM solution of duplex AQ-DNA(1) at 350 nm under the standard conditions followed by treatment of the irradiated sample with piperidine and analysis by PAGE and autoradiography (the standard analytical protocol) shows that strand cleavage occurs at both of the GG steps [12]. Control experiments confirm that this is an intramolecular reaction and is not due to the generation of a diffusible species such as singlet oxygen ($^1O_2$).

The relative amount of strand cleavage at each site of AQ-DNA(1) is indicated by the length of the solid vertical arrow shown in Fig. 4. As is often observed, the 5′-G of the GG steps react more often than do the 3′-G. In the case of AQ-DNA(1), the relative reactivity is ca. 1:3, but this ratio depends upon the specific base pair sequence surrounding a GG step, which may be an indication of radical cation delocalization to bases adjacent to the GG sequence. It is worth pointing out again that these reactions are carried out under single-hit conditions where the relative strand cleavage efficiency seen at various locations of AQ-DNA(1) reflect the statistical probability that the radical cation will be trapped by H_2O at that site.

The results from irradiation of AQ-DNA(1) show conclusively that a radical cation introduced at one site, G_1 at the 3′-terminus of the complementary strand in this case, can migrate through duplex DNA and cause reaction at remote sites. To migrate from its point of injection at G_1 to where it reacts at

GG_8, the radical cation must traverse five A/T base pairs. Electrochemical measurements in solution have shown that the purine bases (A and G) have considerably lower E_{ox} than the pyrimidines (C and T), with the E_{ox} of G estimated to be about 0.25 V below that of A [20]. It is not very likely that the E_{ox} of bases in DNA will be the same as they are in solution, but it is generally assumed that the order of E_{ox} will remain the same. Consequently, the radical cation at G_1 of AQ-DNA(1) must traverse a "bridge" of five A bases to reach GG_8. The process whereby the radical cation crosses such bridges has been a major point of debate in consideration of long distance radical cation migration mechanisms in DNA; this issue will be discussed fully below.

In AQ-DNA(1), GG_8 and GG_{13} are separated by a bridge of three A bases. If GG_8 were a deep trap for the radical cation, then no reaction would be observed at GG_{13}. If the $(A)_3$ bridge separating GG_8 and GG_{13} presented a significant barrier to charge migration, then the amount of strand cleavage at GG_{13} would be significantly less than at GG_8. The experiment reveals that the amounts of reaction at GG_8 and GG_{13} are the same within experimental error, which shows that GG steps are not deep traps and the rate of radical cation migration through an $(A)_3$ bridge is much faster than the reaction of the radical cation with H_2O at either of the two GG steps in this oligomer. More recently, Giese and coworkers have shown radical cation migration through an $(A)_{12}$ bridge [21].

A deep radical cation trap can be introduced into duplex DNA. The E_{ox} of 7,8-dihydro-8-oxoguanine (8-OxoG) is ca. 0.5 V below that of G [22]. Irradiation of an AQ-DNA(1) analog in which an 8-OxoG was substituted for G_8 essentially stops observable strand cleavage at G_7, G_{12}, and G_{13} [12]. In a related series of experiments, irradiation of UAQ-DNA(2), see Fig. 4, under the standard conditions gives strand cleavage at each of the eight GG steps of both strands. But substitution of an 8-OxoG for either of the three guanines on either side of the UAQ, as shown in Fig. 4, results in the reduction of the efficiency of strand cleavage at each G in both strands on the same side of the UAQ as the 8-OxoG [13]. This finding shows that a deep trap will inhibit charge migration both in the strand containing it and in the complementary strand, which demonstrates that the radical cation can cross from one strand of duplex DNA to its complement.

We examined long-distance charge transport in AQ-DNA(3), see Fig. 5, to obtain additional information on how base sequence affects the efficiency of radical cation migration [23]. There are four GG steps in AQ-DNA(3) that are positioned 10, 28, 46, and 55 base pairs from the site of charge injection at G_1. Significantly, there is no regularity of the sequence of bases between any of these GG steps. Irradiation of AQ-DNA(3) under the standard conditions gives detectable strand cleavage at each of the GG steps. The relative amount of strand cleavage at each GG step is indicated by the vertical arrows and is plotted as a semi-log plot against distance in Fig. 5. Remarkably, the radical cation introduced at G_1 migrates nearly 200 Å through "mixed sequence" DNA to cause reaction at G_{55}. Surprisingly, the semilog plot in Fig. 5 reveals an apparent linear relationship between the amount of reaction and the distance between the GG step and the site of charge injection, with

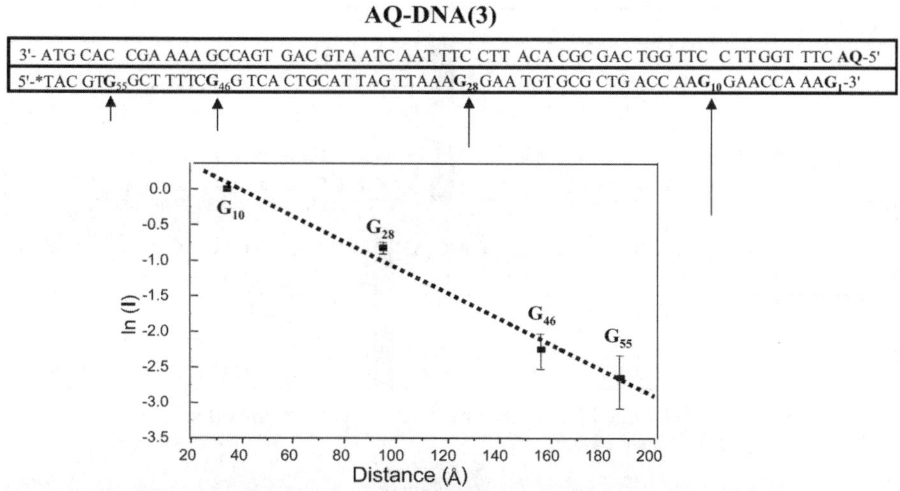

Fig. 5 Schematic representation of long distance radical cation migration in DNA. In AQ-DNA(3), irradiation of the anthraquinone group linked at the 5′-terminus leads to reaction at GG steps that are 10, 28, 46 and 55 base pairs from the charge injection site. The solid arrows indicate approximately the amount of reaction observed at each GG step. The plot shows the natural log of the normalized amount of reaction as a function of distance from the AQ. The results appear to give a linear distance dependence

an exponential distance dependence of ca. -0.02 Å^{-1}, a value that has also been observed with other sequences and with other sensitizers [24]. A linear dependence is unexpected because it requires that the radical cation migrate from base-to-base through both pyrimidine and purine bases or from strand-to-strand with a similar rate constant, independent of the specific order of bases it encounters. There are two reasonable explanations for this observation: either the linear dependence is an artifact; or some process is operating that causes averaging of differences in base E_{ox} that gives a distance dependence which appears to be independent of base sequence.

AQ-DNA(4), see Fig. 6, is related to AQ-DNA(1) – both have a series of GG steps separated by a number of A bases. However, in AQ-DNA(4), there are four GG steps and they are on the AQ-linked strand, which contains only purines and carries the radiolabel at its 3′-terminus. We have shown that the outcome of oxidative reactions of duplex DNA is unaffected by moving the label from one strand to its complement [25]. Irradiation of AQ-DNA(4) under the standard conditions gives the expected outcome. The amount of strand cleavage detected at GG_4 and GG_8 is nearly the same, the $(A)_8$ sequence that separates GG_8 from GG_{18} presents only a modest barrier to the migration of the radical cation: the cleavage efficiency at GG_{18} and GG_{22}, which are approximately equal, is about 40% of the amount detected at GG_4 and GG_8 [26].

The results obtained from irradiation of AQ-DNA(5) are startling in their contrast. This duplex also contains four GG steps, but the (A_8) bridge of

AQ-DNA	
4	5'AQ- A A G G_4 A A G G_8 A A A **A** A A A A G G_{18} A A G G_{22} A A A A* -3' 3'- T T C C T T C C T T T **T** T T T T C C T T C C T T T T -5'
5	5'AQ- A A G G_4 A A G G_8 A A A **T** A A A A G G_{18} A A G G_{22} A A A A* -3' 3'- T T C C T T C C T T T **A** T T T T C C T T C C T T T T -5'
6	5'AQ- A A A T G C C G G T A C C **T** C T A G G C C G T A G -3' T T T A C G G_7 C C A T G G_{13} **A** G A T C C G G_{21} C A T C* -5'
7	5'AQ- A A A T G C C G G T A C C **A** C T A G G C C G T A G -3' T T T A C G G_7 C C A T G G_{13} **T** G A T C C G G_{21} C A T C* -5'

Fig. 6 Structures of AQ-linked DNA oligomers used to assess the effect of converting an A/T base pair to a T/A base pair

AQ-DNA(4) is replaced by an $(A_3)(T)(A_4)$ sequence. In other words, one A/T base pair of AQ-DNA(4) becomes a T/A base pair in AQ-DNA(5). This simple structural change has a profound effect on the efficiency of radical cation transport across the eight-base-pair bridge. As was observed for AQ-DNA(4), GG_4 and GG_8 of AQ-DNA(5) are approximately equally reactive, but the amount of strand cleavage detected at GG_{18} and GG_{22} is reduced by ca. 95% compared with that of GG_4 and GG_8. This is surprising because AQ-DNA(3) has 13 T bases between G_{10} and G_{55} but gives the linear distance relationship that is shown in Fig. 5. Clearly, there is no general principle that requires such a linear distance dependence that is totally independent of base sequence.

In contrast to the overwhelming affect of conversion of an A/T base pair in AQ-DNA(4) to a T/A base pair in AQ-DNA(5) on radical cation transport, the identical change in AQ-DNA(6) and AQ-DNA(7) has no measurable effect on the amount of strand cleavage observed at GG_7 or GG_{21} [27]. It is apparent from consideration of these results that the effect of a change in base sequence must be considered in the context of the surrounding base pairs and not in isolation.

We probed the effect of base sequence on long-distance radical cation migration using a series of duplexes that have a regularly repeating structure of base pairs, see Fig. 7. AQ-DNA(8) can be recognized as containing an AAGG sequence that repeats six times $(AAGG)_6$. The "last" four base pairs of this duplex are $(A/T)_4$, which reduces misalignment of the duplex by slippage. Irradiation of AQ-DNA(8) under the standard conditions gives an essentially equal amount of strand cleavage at each of its six GG steps. This is precisely what is to be expected if the rate of radical cation migration is much faster than the rate of its trapping by reaction with H_2O.

A semi-log plot of the distance dependence of strand cleavage efficiency, see Fig. 8, gives a linear relationship with a slope experimentally indistinguishable from zero. DNA has occasionally been characterized as a "molecu-

AQ-DNA	
8	5'AQ- A A GG_4 A A GG_8 A A GG_{12} A A G G_{16} A A GG_{20} A A GG_{24} A A A A* -3' 3'- T T C C T T C C T T C C T T C C T T C C T T C C T T T T -5'
9	5'AQ- A T GG_4 A T GG_8 A T GG_{12} A T G G_{16} A T GG_{20} A T GG_{24} A T A T* -3' 3'- T A C C T A C C T A C C T A C C T A C C T A C C T A T A -5'
10	5'AQ- A T A G G_5 A T A GG_{10} A T A GG_{15} A T A GG_{20} A T A T* -3' T A T C C T A T C C T A T C C T A T C C T A T A -5'
11	5'AQ- A T T A G G_6 A T T A GG_{12} A T T A GG_{18} A T T A GG_{24} A T A T* -3' T A A T C C T A A T C C T A A T C C T A A T C C T A T A -5'

Fig. 7 Structures of AQ-linked DNA oligomers containing a regularly repeating sequence of base pairs that were used to assess the effect of base sequence effects on long-distance radical cation migration

lar wire" [28]; a phrase that lacks a precise definition. The behavior of AQ-DNA(8) is the most "wire-like" that has been reported, but this does not qualify it as a molecular wire. This point will be addressed more extensively below.

It is especially informative to compare the behavior of AQ-DNA(9) with that of AQ-DNA(4) and AQ-DNA(8). AQ-DNA(9) contains the repeating sequence $(ATGG)_6$, which can be thought of as being formed from AQ-DNA(8) by converting the A/T base pair preceding each GG step to a T/A base pair. Recall that one such change converted AQ-DNA(4) to AQ-DNA(5) and resulted in the introduction of a high barrier to radical cation migration across the $(A_3)(T)(A_4)$ bridge that this change created. In contrast, radical cation migration through the five T/A base pairs between GG_1 and GG_{24} of

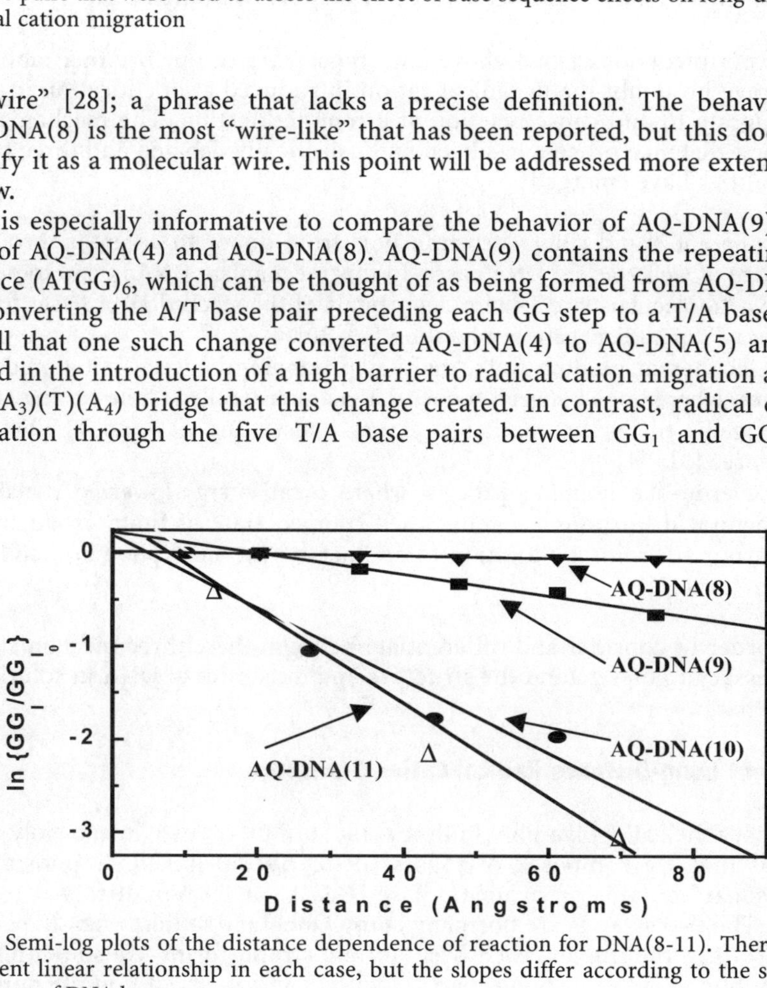

Fig. 8 Semi-log plots of the distance dependence of reaction for DNA(8-11). There is an apparent linear relationship in each case, but the slopes differ according to the specific sequence of DNA bases

AQ-DNA(9) is hardly affected. The slope of the line shown in Fig. 8 for AQ-DNA(9) is –0.008 Å^{-1}, which shows that the five T/A base pairs combined result in only a ca. 50% reduction in radical cation transport efficiency from GG_1 to GG_{24}, whereas the single T/A base pair of AQ-DNA(5) causes a 95% reduction in the efficiency of radical cation migration from GG_8 to GG_{18}. It is a similar case for AQ-DNA(10) and AQ-DNA(11), where one and two T/A base pairs are interposed between GG steps, respectively, with only a modest affect on radical cation migration from GG to GG. These observations show that the effect of base sequence on radical cation migration cannot be analyzed by considering the base pairs in isolation; base-to-base charge interaction evidently plays a key role.

5 Mechanisms of Long-Distance Charge Transport in Duplex DNA

The experiments described above, and those carried out in other laboratories, leave no doubt that a radical cation introduced at one location in DNA can migrate to and cause reaction at a remote location. The mechanism of this long-distance process has been enthusiastically debated and three broad possibilities have emerged:

a. A coherent, rapid single-step transport from donor to acceptor through a bridge of well-stacked DNA bases. In this mechanism DNA is said to behave like a "molecular wire" where the orbitals of the stacked DNA bases form a "π-way" for radical cation migration [29, 30].
b. An incoherent random-walk, multi-step hopping between initial and final states, where hops between sequential guanines (called "hole resting sites") are mediated by superexchange across intervening A/T and T/A base sequences [31–34].
c. A polaron-like hopping process where local energy-lowering dynamical structural distortions generate a self-trapped state of finite extent that is transported from one location to another by thermal (phonon) activation [7, 23, 35–37].

In order to consider and differentiate between these three mechanisms, it is necessary to understand the structure and dynamics of DNA in solution.

6 Coherent Long-Distance Radical Cation Transport

DNA is a helical polyanion built by the union of two linear polymeric strands that are composed of sugars (deoxyribose) linked by phosphates. Each sugar contains an aromatic base (G,C,A, or T) bound to C-1′ of the sugar. The two strands are normally complementary so that when they combine to form the duplex, each base on one strand forms Watson-Crick hydrogen bonds with its counterpart (G with C and A with T) on the opposite strand. At normal physiological pH (ca. 7.4), the phosphates of the backbone polymer are fully ionized, so there must be a counterion (Na^+) for each phosphate. In fact, duplex DNA is unstable in solutions of low ionic strength because of Coulombic repulsion of the phosphate anions that is normally screened by the counterions [38].

High-resolution X-ray crystallography of DNA reveals exquisite details about its structure. In B-form DNA, the medium most commonly used for the study of long-distance radical cation transport in solution, the average distance from one base pair to the next is 3.4 Å, and each base pair is rotated around the long axis of the helix by about 36° with respect to its adjacent base pairs [39]. The regular order of stacked bases revealed by this structure led naturally to the suggestion that DNA was able to support long-distance electron transport [40]. This exciting possibility was revived and supported by measurements of apparent rapid photoinduced charge transfer over more than 40 Å between metallointercalators tethered to opposing 5′-termini of a 15 base pair DNA duplex [29, 30].

However, careful kinetic measurements on related systems showed the invalidity of wire-type behavior [41]. Furthermore, Sen and coworkers [42] recently showed that the appearance of rapid, long-distance charge transfer for metallointercalators may be an artifact caused by the formation of aggregates. Currently, there are no data that clearly support the existence of a coherent transfer process in DNA over a distance greater than one or two base pairs [43, 44].

The crystallographic structure of DNA is not a good model for consideration of the possibility that it behaves like a "molecular wire" in solution because this structure does not reveal the extent of instantaneous disorder inherent in this assembly. DNA is a dynamic molecule with motions of its constituent atoms, corresponding counterions and solvating water molecules that occur on time scales that range from femtoseconds to milliseconds or more. This is revealed clearly by consideration of careful molecular dynamics simulations [45]. It is apparent from analysis of these simulations that duplex DNA in solution has the standard B-form structure on average, but at any instant, over long distances (more than three or four base pairs) the DNA is somewhat disordered. Disorder cannot be tolerated in a coherent, single-step charge transfer process because it greatly reduces the electronic interaction that couples one base pair to the next [46]. Consequently, DNA in solution cannot be a molecular wire and this mechanistic possibility must be discarded.

7
Hopping Models: Hole-Resting-Site and Phonon-Assisted Polaron Transport

It is now clearly demonstrated that a radical cation introduced at one location in duplex DNA can migrate 200 Å or more and result in reaction at a remote GG step. Consideration of the dynamical nature of DNA in solution led to the suggestion that this long-distance migration was the result of a radical cation hopping process [47]. In this view, the radical cation is trapped in a shallow minimum localized on a single base, or delocalized over several bases, and some process causes it to move from one location to the next until it is finally trapped irreversibly by reaction with H_2O.

In one variant of the charge-hopping mechanism, called the *hole-resting-site model*, the radical cation is localized on individual guanines and tunnels through bridges composed of A/T and T/A bases from strand-to-strand until it is trapped. Although this was considered to be a general process when it was first suggested, now it is viewed to be valid only for bridges containing three or fewer base pairs [34].

In a second possibility, the *polaron-like hopping model*, a structural distortion of the DNA stabilizes and delocalizes the radical cation over several bases. Migration of the charge occurs by thermal motions of the DNA and its environment when bases are added to or removed from the polaron [23].

The key differences between these representations is that in the hole-resting-site model, the radical cation is localized and confined to guanines, and migrates by tunneling through orbitals of the bridging A/T bases without ever residing on the bridge: the radical cation exists only virtually on the bridge. In the polaron-like hopping representation, the radical cation resides briefly as a real, measurable physical entity on the bridging bases and its hopping occurs by thermal activation.

The hole-resting-site and polaron-like hopping models can be distinguished by the distance and sequence behavior of radical cation migration. Analysis of the hole-resting-site model leads to the prediction that the efficiency of radical cation migration will drop ca. ten-fold for each A/T base pair that separates the G resting sites [33].

This possibility was explored experimentally by investigating the reactions of the DNA oligomers shown in Fig. 9 [19]. In AQ-DNA(12), the oligomer contains a series of six GG steps that are separated by TT sequences. In AQ-DNA(13) through AQ-DNA(15), the GG steps in related oligomers are separated by TTT, TTTT, and TTTTT sequences, respectively. Irradiation of

AQ-DNA	
12	5'-*GG$_7$CC GG$_6$TT GG$_5$CC GG$_4$TT GG$_3$CC GG$_2$TT GG$_1$CCAAAA-3' 3'-CC$_7$GG CC$_6$ AA CC$_5$GG CC$_4$AA CC$_3$GG CC$_2$ AA CC$_1$GGTTTT-AQ-5'
13	5'-*GG$_7$CC GG$_6$TTT GG$_5$CC GG$_4$TTT GG$_3$CC GG$_2$TTT GG$_1$CCAAAA-3' 3'-CC$_7$GG CC$_6$ AAA CC$_5$GG CC$_4$AAA CC$_3$GG CC$_2$ AAA CC$_1$GGTTTT-AQ-5'
14	5'-*GG$_7$CC GG$_6$TTTT GG$_5$CC GG$_4$TTTT GG$_3$CC GG$_2$TTTT GG$_1$CCAAAA-3' 3'-CC$_7$GG CC$_6$ AAAA CC$_5$GG CC$_4$AAAA CC$_3$GG CC$_2$ AAAA CC$_1$GGTTTT-AQ-5'
15	5'-*GG$_7$CC GG$_6$TTTTT GG$_5$CC GG$_4$TTTTT GG$_3$CC GG$_2$TTTTT GG$_1$CCAAAA-3' 3'-CC$_7$GG CC$_6$ AAAAA CC$_5$GG CC$_4$AAAAA CC$_3$GG CC$_2$AAAAA CC$_1$GGTTTT-AQ5'

Fig. 9 Structures of AQ-linked DNA oligomers used to assess the effect of $(T)_n$ sequences between GG steps

these assemblies under standard conditions, and examination of the effect of sequence on strand cleavage yields at the GG steps gives in each case a semi-log plot, with a linear distance dependence having a slope within experimental error of $-0.02\ \text{Å}^{-1}$. This value corresponds to only a ca. 10% reduction in radical cation transport efficiency for each intervening T base, which is inconsistent with the prediction of the hole resting site model. These experimental results, in part, led to the current view that tunneling from G to G cannot compete with other processes if the guanines are separated by more than three base pairs [34].

Support for the hole-resting-site model is built on the assumption that the radical cation migrates through a lattice of base pairs frozen in the standard B-form structure of DNA. However, in solution at room temperature at any given instant, only very short segments of the oligomer have their bases at precisely the B-form locations. Moreover, the magnitude of the electronic coupling interaction between adjacent bases is very strongly dependent on the details of the instantaneous structure [46]. Consequently, the factorization of the radical cation transport rate into an electronic coupling term and one due to nuclear vibrational motion (a Franck-Condon factor), employed for quantitative interpretation of the hole-resting-site model, does not apply to long distance migration of radical cations in DNA in solution.

However, such a process might operate over short distances where radical cation migration is forced to occur on a short time scale by a rapid back electron transfer reaction [43]. In such a circumstance, tunneling from G to G may occur in those DNA molecules from among the entire ensemble of molecules that happen to have structures permitting strong electronic coupling between relevant base pairs at the instant the radical cation is created. The dynamical structure of DNA in solution guarantees that such an arrangement can extend for no more than a very few base pairs, and perhaps occurs only when the DNA is constrained in a relatively rigid structure such as a hairpin [43]. On this basis, the hole-resting-site model cannot be the entire explanation for the observation of radical cation migration of 200 Å or more in duplex DNA.

The phonon-assisted polaron-like hopping model is unique because it is built upon an understanding of the dynamical nature of DNA in solution. The fundamental assumption of this model is that the introduction of a base radical cation into DNA will be accompanied by a consequent structural change that lowers the energy for the system.

A base radical cation is a highly electron-deficient species: it will be stabilized and the energy of the system will be reduced by changes in the average orientations of nearby bases, counterions and solvent molecules that provide additional electron density to the radical cation. This process, of course, will delocalize the radical cation and cause a local distortion of the DNA structure so that, on average, it is no longer in the standard B-form. This distortion may not extend over very many base pairs because the stabilization gained by delocalization must be balanced by the energy required to distort the average DNA structure. In this view, radical cations in DNA are self-trapped species that are delocalized over several base pairs contained within

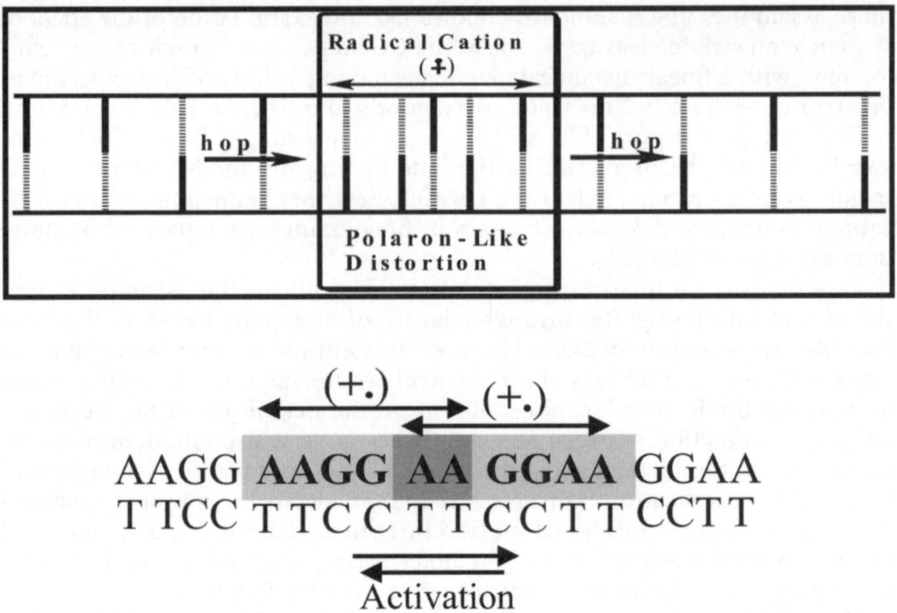

Fig. 10 Two schematic representations of a polaron-like species in DNA. In the top drawing, the base pairs of DNA are represented by the horizontal lines; the sugar diphosphate backbone is represented by the vertical lines. The polaronic distortion is enclosed in the box and extends over some number of base pairs. This is shown schematically by drawing the base-pair lines closer together. In the lower figure, a specific potential polaron is identified, AAGGAA, and the radical cation is presented as being delocalized over this sequence. Movement of the polaron from one AAGGAA sequence to the next requires thermal activation

a distorted local structure, which is the definition of a small polaron [48]. In fact, a base radical cation in DNA is more precisely referred to as a "polaron-like" species, because for most DNA oligomers the sequence of base pairs does not follow any particular repeating rule that would allow the classical polaron behavior that is observed in one-dimensional conducting polymers, for example [49].

Figure 10 shows schematic representations of possible polaron-like species in DNA. In the upper part of the Figure, the DNA bases are represented as a series of vertical lines (dashed for the purines and solid for the pyrimidines) distributed between horizontal lines that represent the sugar-diphosphate backbone. The box within this representation portrays the distortion of the polaron by placing the base pairs closer together in this region. This distortion of the DNA structure from its normal B-form average results in the delocalization of the radical cation (probably unevenly) among the bases included in the distortion. The polaron-like distortion is considered to hop through the DNA duplex, a process that may either increase or reduce the number of base pairs in the polaron; the size of the hopping step will be controlled by the sequence of bases that make up the polaron and by those surrounding it.

The polaron-hopping model accommodates the experimental data obtained from long-distance radical cation migration experiments. For example, the apparent linear relationship between the log of the cleavage efficiency and distance observed for AQ-DNA(3), shown in Fig. 5, can be explained qualitatively by supposing that polaron-hopping permits two kinds of averaging that tend to reduce the effect of specific base sequence on radical cation migration efficiency. The observed linear relationship implies that the barrier for each hopping step the polaron takes is of approximately the same height, independent of specific base sequence. The height of the barrier is the difference between the energy of a polaron and the transition state that separates one polaron from the next.

The stabilization of the radical cation by forming a polaron is a trade-off between its delocalization and the energy required to distort the DNA structure. The former lowers the kinetic energy of the intrinsically quantum mechanical migrating radical cation, and the latter will be determined by factors that are independent of specific base sequence, such as the force constants of bonds in the sugar diphosphate backbone.

For example, if a strand of DNA is composed of sequential adenines or guanines (A_n or G_n), comparable stabilization of the polaron would likely involve fewer bases than in a segment having a mixed sequence of purines and pyrimidines. However, the relative energies of the two polarons could be averaged to a similar value even though they extend over a different number of bases having different sequences. The energy of the transition state that separates two polarons may also become less dependent on specific base sequence by averaging. There is no requirement that the number of bases in a hop from one location to the next be constant. If the hopping length is somehow dependent upon the identity of the bases separating the polarons, the energy of the transition state may depend less on the base sequence. Thus, the energy of the polaron is averaged over several bases and the energy of the transition state is averaged by different hopping lengths. The postulation that polaron formation accounts for the observed linear distance dependence of AQ-DNA(3) and similar experiments is qualitative. Polaron formation can be placed on a firmer footing by consideration of the experiments with AQ-DNA(4) through AQ-DNA(11).

8 Base Sequence Effects on Radical Cation Migration in DNA – A Collective Phenomenon

The linear distance dependence seen for AQ-DNA(3) is not observed to be universally independent of specific DNA base sequence. This is clearly revealed by examination of AQ-DNA(4) and AQ-DNA(5). Plots of the distance dependence of strand cleavage at the GG steps in these oligomers are shown in Fig. 11. Both show “stepped” rather than linear behavior, and the size of

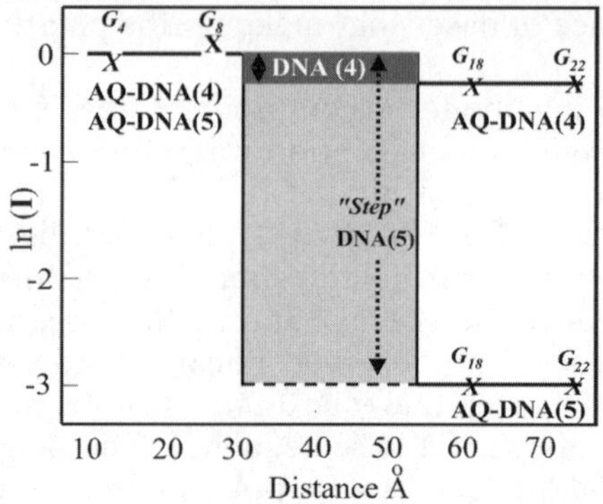

Fig. 11 Semi-log plots of the distance dependence of the reactivity of AQ-DNA(4) and AQ-DNA(5). These oligomers show "stepped" rather than linear behavior. The size of the step is strongly dependent on the details of the structure

the step is dramatically dependent on the base sequence. In both of these assemblies, the amount of strand cleavage at G_4 and G_8 is approximately equal, but amount of strand cleavage at G_{18} and G_{22} is reduced (the step), and the size of the step for AQ-DNA(4) is much less than it is for AQ-DNA(5). Clearly, averaging by polaron formation is not sufficient to give a linear distance dependence for these two sequences.

Further insight into polaron formation and sequence averaging comes from consideration of AQ-DNA(8), Fig. 7, which shows a linear distance dependence with a slope close to zero, Fig. 8. A slope of zero means that every GG step regardless of its distance from the AQ (the site of charge injection) reacts with the same efficiency. The kinetic model presented above reveals that this behavior is expected when the rate of radical cation migration is much faster than the rate of its irreversible trapping with H_2O. Since the rate of the trapping reaction is considered to be constant, a slope of zero suggests that the barriers to migration of the radical cation are significantly reduced in AQ-DNA(8) compared with AQ-DNA(3), which, for example, has a slope of -0.02 Å^{-1}. Figure 10 presents an explanation for this behavior based on the arbitrary assignment of the polaron in AQ-DNA(8) to the AAGGAA sequence.

In this formulation, the polaron is specially stabilized by the AAGGAA sequence, and identical polarons are separated by an AA sequence, which is presumed to present a relatively low-energy transition state that is easily overcome by thermal activation. This proposal is shown graphically in Fig. 12 where a potential energy surface for hopping of the {AAGGAA} polaron over an [AA] barrier ($\Delta G_8^{\neq}$) is qualitatively sketched. AQ-DNA(5) also has the AAGGAA sequence of bases and we similarly assign a specially stabilized polaron in this case. However, unlike AQ-DNA(8), the transition state between the polaron in AQ-DNA(5) centered on GG_8 and the one centered on GG_{18} contains an ATA sequence, which in this case appears to present a nearly insurmountable barrier to radical cation migration ($\Delta G_5^{\neq}$).

Having an ATA sequence between assigned polarons does not always create a high barrier for radical cation migration. In AQ-DNA(10), we assign

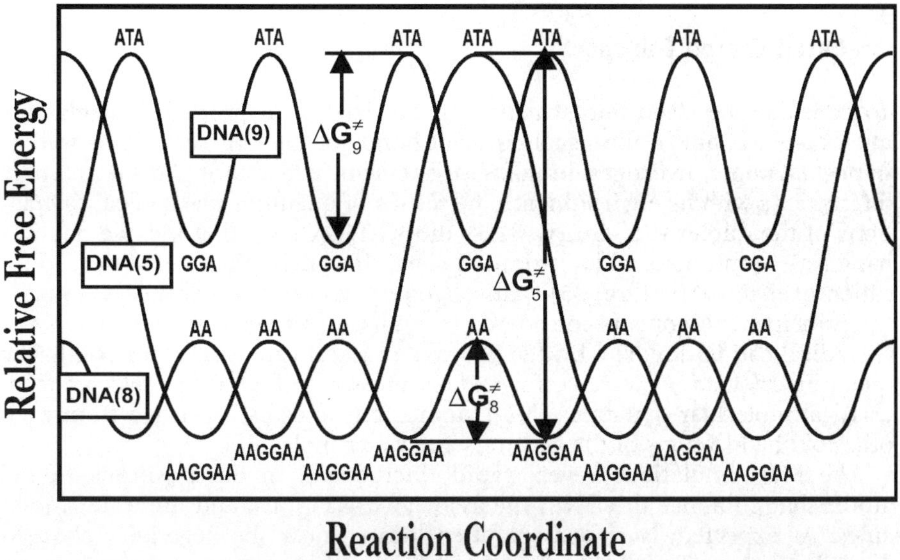

Fig. 12 A reaction coordinate diagram illustrating the emergence of sequence effects in long distance charge transport in duplex DNA. The curve representing DNA(8) shows the radical cation delocalized and stabilized in polarons; identified arbitrarily here as AAGGAA sequences in the AAGGAAGGAA segments surrounding the ATA sequence. This delocalization of the radical cation stabilizes it and results in a high barrier ($\Delta G^{\neq}{}_5$) at the ATA sequence; trapping of the radical cation by water occurs much faster than this barrier can be crossed. For DNA(8), the same AAGGAA polaron is identified and there are no thymines that create a high barrier for hopping from one polaron to the next, which occurs faster than trapping by water. The curve that represents DNA(9) shows an intermediate case where the polaron is assumed to be the GGA sequence, which is less delocalized and therefore higher in energy than AAGGAA. Consequently, the barrier introduced by the ATA sequence ($\Delta G^{\neq}{}_9$) is lower than for DNA(5) and the rate of crossing this barrier is comparable with reaction of the radical cation with water

the polaron to the {AGGA} sequence, because it is bracketed by T bases. Further, we presume that the {AAGGAA} polaron, being more delocalized, has a lower relative energy than the {AGGA} polaron. Consequently, the barrier to charge migration for the {AGGA} polaron when it encounters an [ATA] transition state ($\Delta G_9{}^{\neq}$)is lower than when the {AAGGAA} polaron encounters the same transition state sequence. This proposal is also illustrated in Fig. 12.

The primary conclusion that follows from the effect of base sequence on the efficiency of radical cation migration through duplex DNA is that base pairs cannot be considered in isolation. For example, the effect of placing a T in a sequence of purines depends critically on the nature and number of purines. In this regard, the effect of base sequence on radical cation transport emerges from examination of collective properties of the DNA. This is a clear indication that the charge is delocalized over several base pairs, a conclusion that is supported by extensive quantum calculations.

9
Ion-Gated Charge Transport

To consider the electronic structure of oxidized DNA properly, calculations must take account of the usual covalent bonds of the double helix as well as important ionic, hydrogen bonding, dispersion, and multipolar electrostatic interactions with its environment. The results of quantum-mechanical calculations of the duplex d(5′-$G_1A_2G_3G_4$-3′)·d(3′-$C_5T_6C_7C_8$-5′), that include neutralizing Na^+ counterions and a hydration shell, show delocalization of the radical cation over this structure [35]. This oligomer was selected because it contains the principal components considered in studies of charge transport in DNA: a G (radical cation donor) a bridge (A) and a radical cation acceptor (GG). The quantum calculations were performed on nuclear configurations selected from classical molecular dynamics (MD) simulations and distinguished from each other by the locations of the Na^+ ions and water molecules.

The MD simulations reveal rapid fluctuations in the positions of the atoms that compose the DNA, the associated Na^+ ions, and the water molecules. As expected, Na^+ ions are often located near the negatively charged phosphate groups of the backbone and near the relatively electronegative atoms (N-7 of G and A, for example) of the bases [45, 50].

Results obtained from the quantum calculations for configurations of the native and ionized duplex with the Na^+ ions near the phosphate groups are shown in Fig. 13. The highest occupied molecular orbital of the native DNA is, as expected, found to be on the G_3G_4 step. However, surprisingly, the radical cation is delocalized over this sequence with major density on G_1 and on the G_3G_4 step, and with a small amount of charge on A_2. The vertical ionization potential calculated for the d(5′-$G_1A_2G_3G_4$-3′)·d(3′-$C_5T_6C_7C_8$-5′) duplex with this configuration of Na^+ ions and solvating water molecules is 5.22 eV.

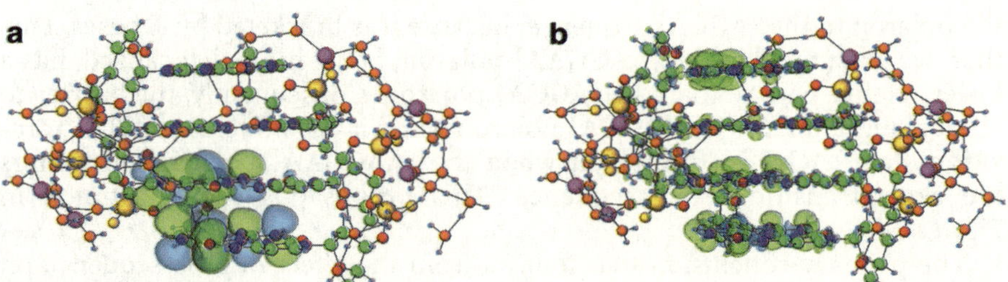

Fig. 13 Results from the quantum calculations on the duplex sequence 5′-GAGG-3′. In **a**, the sodium ions and their solvating water molecules are located at positions near the phosphate anions of the DNA backbone. In **b**, one sodium ion is moved from near a phosphate anion to N-7 of a guanine, which molecular dynamics calculations show to be a preferred site. The "balloons" represent the hole density on the GAGG sequences with the two different sodium ion orientations. The radical cation clearly changes its average location with movement of the sodium ion

Significantly, changing the position of only one Na^+ ion from near the phosphate group linking G_3 and G_4 to a favored position near N-7 of G_4 reduces the radical cation density at the G_3G_4 step, and raises the vertical ionization potential of the duplex to 5.46 eV. Even more revealing is the relocation of the Na^+ ion found at the phosphate group linking A_2 and G_1 to N-7 of G_1. In this case, movement of the single Na^+ ion causes the radical cation to localize on the G_3G_4 step and the calculated vertical ionization potential of the duplex to increase to 5.69 eV. It is important to note that the magnitude of the fluctuation in vertical ionization potential caused by the relocation of just one Na^+ ion (and its accompanying water molecules) is greater than the measured difference in ionization potential between a G (the hole donor) and an A (the "bridge"). These findings indicate that thermal fluctuations of Na^+ ions allows the system to access a configuration in which the energy of the "bridge state" is below the energy of the hole donor.

These calculations show that a radical cation in DNA is delocalized and that its motion through the duplex is controlled, at least in part, by the motions of the Na^+ ions by a process we describe as *ion-gated charge transport*. In the ion-gated transport model, a radical cation (which may extend over several DNA bases) hops from one location to another by transitions between (quantum mechanical) states that are governed by the dynamically evolving local configurations of the Na^+ ions and water molecules. This concept is represented pictorially in Fig. 14 where the energy of the radical cat-

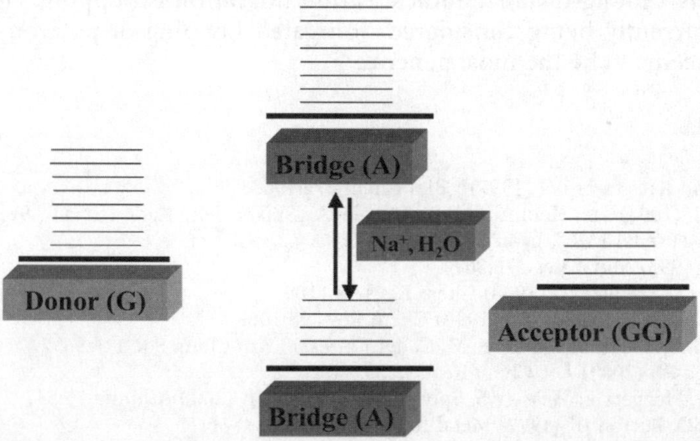

Fig. 14 Schematic representation of the ion-gated radical cation transfer postulate. A radical cation at the "donor site", identified as an isolated G, migrates to the "acceptor site", a GG step, through a bridge composed of contiguous A bases. The energy of the bridge is modulated by movements of the sodium ions and their accompanying water molecules. When the energy of the bridge comes close to the energy of the hole on the donor, the hole hops onto the bridge. Further motions result in additional energy changes that can cause the hole to migrate from the bridge to the acceptor. Of course, motions of the sodium ions can also modulate the energies of the hole donor and acceptor, but since only relative energies are relevant, these two possibilities are operationally equivalent

ion donor (G) and the radical cation acceptor (GG) are above or below the energy of the A bridge, which itself depends on the Na^+ ion configuration. In this model, there is no radical cation tunneling from G to GG through the A, and the rate determining step for the radical cation to hop is controlled by the global structure of the DNA and its environment.

In some conformations of the atoms that compose the DNA, the Na^+ ions, and the water molecules, the energy of the system with radical cation density on the bridging A is below that of configurations where the radical cation is localized on the G or GG, and it is under these conditions that the radical cation hops to the A where it resides briefly. At some nuclear configurations of the system, the energy of the radical cation on the GG step is below its energy on the bridging A or the donor G. These configurations are more likely to occur than those that stabilize the radical cation on the A or G; and consequently, the radical cation remains on the GG step for a longer time, which permits it to be trapped occasionally by H_2O.

10 Conclusions

Oxidation of DNA (loss of an electron) generates a radical cation that can migrate long distances to remote guanines in G_n steps where it is trapped by H_2O. Irradiation of anthraquinone-linked DNA oligomers is an efficient and effective method for introducing a radical cation into duplex DNA. The mechanism of long-distance radical cation migration is hopping. Of the two models currently being considered, ion-gated hopping of polaron-like distortions seems to be the most general.

References

1. Beckman KB, Ames BN (1997) J Biol Chem 272:19633
2. Cadet J (1994) In: Hemminiki K, Dipple A, Shiker DE, Kadlubar FF, Segerback D, Bartsch H (eds) IARC, Lyon
3. Sies H (1993) Mutat Res 275:367
4. Burrows CJ, Muller JG (1998) Chem Revs 98:1109
5. Pogozelski WK, Tullius TD (1998) Chem Revs 98:1089
6. Kasai H, Yamaizumi Z, Berger M, Cadet J (1992) J Am Chem Soc 114:9692
7. Schuster GB (2000) Acc Chem Res 33:253
8. Giese B, Meggers E, Wessely S, Spormann B, Biland A (2000) Chimia 54:547
9. Kelley SO, Barton JK (1999) Metal Ions Biological Syst 36:211
10. Rehm D, Weller A (1970) Israel J Chem 8:259
11. Armitage BA, Yu C, Devadoss C, Schuster GB (1994) J Am Chem Soc 116:9847
12. Gasper SM, Schuster GB (1997) J Am Chem Soc 119:12762
13. Ly D, Sanii L, Schuster GB (1999) J Am Chem Soc ??:9400
14. Sanii L, Schuster GB (2000) J Am Chem Soc 122:11545
15. Deshmukh H, Joglekar SP, Broom AD (1995) Bioconjugate Chem 6:578
16. Nakatani K, Dohno C, Saito I (1999) J Am Chem Soc 121:10854
17. Saito I, Takayama M, Sugiyama H, Nakatani K, Tsuchida A, Yamamoto M (1995) J Am Chem Soc 117:6406
18. Prat F, Houk KN, Foote CS (1998) J Am Chem Soc 120:845

19. Sartor V, Boone E, Schuster GB (2001) J Phys Chem B 105:11057
20. Steenken S, Jovanovic SV (1997) J Am Chem Soc 119:617
21. Giese B, Biland A (2002) Chem Commun ??:667
22. Hickerson RP, Prat F, Muller JG, Foote CS, Burrows CJ (1999) J Am Chem Soc 121:9423
23. Henderson PT, Jones D, Hampikian G, Kan Y, Schuster GB (1999) Proc Natl Acad Sci USA 96:8353
24. Nunez M, Hall DB, Barton JK (1999) Chemistry & Biology 6:85
25. Santhosh U, Schuster GB (2002) J Am Chem Soc 124:10986
26. Liu C-S, Schuster GB (2003) J Am Chem Soc (submitted for publication)
27. Barnett RN, Cleveland CL, Landman U, Boone E, Kanvah S, Schuster GB (2003) J Phys Chem B (in press)
28. Stemp EDA, Barton JK (1996) Metal Ions in Biol Systems 33:325
29. Murphy CJ, Arkin MR, Jenkins Y, Ghatlia ND, Bossman SH, Turro NJ, Barton JK (1993) Science 262:1025
30. Turro NJ, Barton JK (1998) J Biol Inorg Chem 3:201
31. Jortner J, Bixon M, Langenbacher T, Michel-Beyerle ME (1998) Proc Natl Acad Sci USA 95:12759
32. Bixon M, Giese B, Wessely S, Langenbacher T, Michel-Beyerle ME, Jortner J (1999) Proced Natl Acad Sci USA 96:11713
33. Bixon M, Jortner J (2000) J Phys Chem B 104:3906
34. Jortner J, Bixon M, Voityuk AA, Rosch N (2002) J Phys Chem A 106:7599
35. Barnett RN, Cleveland CL, Joy A, Landman U, Schuster GB (2001) Science 294:567
36. Rakhamanova SV, Conwell EM (2001) J Chem Phys B 105:2056
37. Conwell EM, Rakhmanova SV (2000) Proc Natl Acad Sci USA 97:4556
38. Bloomfield VA, Crothers DM, Tinoco JI (1999) Nucleic Acids: Structure, Properties, and Function. University Science Books, Sausalito
39. Dickerson RE (1992) Methods in Enzymol 211:67
40. Eley DD, Spivey DI (1962) Trans Farad Soc 58:411
41. Wan CZ, Fiebig T, Schiemann O, Barton JK, Zewail AH (2000) Proc Natl Acad Sci USA 97:14052
42. Fahlman RP, Sharma RD, Sen D (2002) J Am Chem Soc 124:ASAP (??)
43. Lewis FD, Zuo X, Hayes RT, Wasielewski MR (2002) J Am Chem Soc 124:4568
44. Shafirovich V, Dourandin A, Huang WD, Luneva NP, Geacintov NE (1999) J Phys Chem B 103:10924
45. Beveridge DL, McConnel KJ (2000) Current Opin Struct Biol 10:182
46. Troisi A, Giorgio Orlandi G (2002) J Chem Phys B 106:2093
47. Ly D, Kan Y, Armitage B, Schuster GB (1996) J Am Chem Soc 118:8747
48. Sewell GL (1962) Polarons and Excitions. Plenum Press, New York
49. Emin D (1986) Handbook of Conducting Polymers. Marcel Dekker, New York
50. McFail-Isom L, Sines CC, Williams LL (1999) Curr Opin Struct Biol 9:298

Top Curr Chem (2004) 236:163–186
DOI 10.1007/b94415

Charge Transport in Duplex DNA Containing Modified Nucleotide Bases

Kazuhiko Nakatani · Isao Saito

Department of Synthetic Chemistry and Biological Chemistry, Faculty of Engineering, Kyoto University, 615-8510 Kyoto, Japan
E-mail: nakatani@sbchem.kyoto-u.ac.jp
E-mail: saito@sbchem.kyoto-u.ac.jp

1 Introduction

DNA-mediated charge transport has been a hot topic in chemistry for the last decade. Studies of DNA-mediated guanine radical cation (hole) transport are often focused on the efficiency of the transport between two guanines separated by a large distance, through the π-stack of DNA base pairs [1–14]. The efficiency of the hole transport could be easily determined if one knew the distance between the hole donor and acceptor, the relative number of holes that reach the acceptor, and the time required for the transport.

The hole transport can be divided into three phases: hole injection, hole transport, and hole trapping (the sites of hole injection and trapping are used for determining the distance of hole transport). A GG doublet having a lower ionization potential than that of a single isolated guanine has been conventionally used for the hole trapping device [15–17].

In contrast, hole injection into the DNA π-stack is not as straighforward as a hole trapping process. In order to inject a hole site-selectively, a variety of oxidizing agents such as Rh(III) intercalator [8–9], anthraquinone derivative [10], glycosyl radical [11, 12], and stilbene [13] are incorporated into DNA at predetermined sites. Since the chemical properties of these oxidizing agents and the mode of incorporation into DNA vary, the chemistry of hole injection is diverse, and it is this diversity between experimental systems that has made the direct comparison of the results in different laboratories more difficult and caused much controversy about the mechanism of DNA mediated charge transport. We have developed a novel electron-accepting nucleoside containing a *p*-cyanobenzophenone substituted uridine ($d^{CNBP}U$) [1–7, 18]. Photoirradiation of oligomers containing $d^{CNBP}U$ in the CX/AG and GTX/AAC (X=$d^{CNBP}U$) sequences effectively produces the hole at the G in the sequence (Fig. 1).

With the site-selective hole injection and the hole trapping device established, the efficiency of the hole transport between the hole donor and acceptor, especially with respect to the distance and sequence dependence, were examined. Our experiments showed that hole transport between two guanines was extremely inefficient when the intervening sequence consisted of more than 5 A-T base pairs [1]. Hole injection into the DNA π-stack using photoexcited $d^{CNBP}U$ was accompanied by the formation of $d^{CNBP}U$ anion radical. Therefore, hole transport would always compete with the back electron transfer (BET). To minimize the effect of BET, we opted for hole transport between G triplets, that are still lower in oxidation potential than G doublet. With this experimental system, we researched the effect of the bridging sequence between two G triplets on the efficiency of hole transport [2].

In a real biological system, DNA is mostly surrounded by many proteins. Protein binding to DNA involves a number of hydrogen bonds and electrostatic contacts between two biopolymers, and induces not only structural deviation from the typical B-form structure, but also electronic perturbation of the π-stacked array of base pairs. We tackled the electronic effects of protein binding on the efficiency of hole transport by using a restriction en-

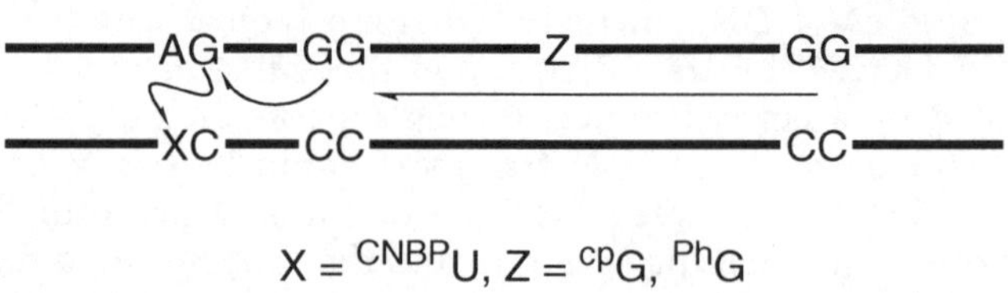

Fig. 1 An image of DNA-mediated charge transport. **X** and **Z** are a modified nucleotide bases

zyme, *Bam*H I, that binds to DNA by hydrogen bonding between guanidium and the G in the 5′ side of the recognition sequence of GGATCC. *Bam*H I binding effectively terminates hole transport beyond the binding site [5].

The G doublet and triplet effectively functions as a thermodynamic sink in DNA-mediated hole transport. However, the rate determining step of hole trapping at guanine clusters and the rate of hole trapping are not well understood. Furthermore, hole transfer between the donor and acceptor should compete with the hole trapping reaction; the relative rate of hole trapping versus the hole transfer rate determines the overall efficiency for the hole transfer.

Therefore, two modified guanines have been designed to trap the hole much faster than the GG doublet does. One modified guanine contains a cyclopropyl group on the amino group at the C-2 position [4]. The cyclopropyl group was expected to undergo rapid ring opening upon formation of the radical cation. The other guanine involves a phenyl group on its amino group [6]. In the phG-containing oligomers, the hole was annihilated without causing irreversible degradation of not only phG but also the "conjugate guanines" to the modified base.

2 Site-Selective Hole Injection by Cyanobenzophenone Substituted Uridine

2.1 Hole Injection in the d(C^{CNBP}U)/d(AG) Sequence [1]

We have developed ODNs that contain a strong electron accepting chromophore at predetermined sites without perturbing the base stack in B-form duplex by incorporating cyanobenzophenone substituted 2′-deoxyuridine (d^{CNBP}U) [18]. The photoreactions of a series of oligomer duplexes containing both d^{CNBP}U and GG hole trap which were separated by various intervening base sequences were examined. Photoreactions were carried out with a variety of duplexes consisting of d^{CNBP}U-containing 22-mer ODNs and their complementary strands. The GG step was incorporated into GGTTG*A* (ODN **1**), GGTTGT*A* (ODN **3**), *A*GTTGG (ODN **5**), and *A*TGTTGG (ODN **7**) sequences, where A (shown in italic) forms a Watson-Crick base pair with d^{CNBP}U (Table 1). Photocleavage sites of ODNs **1**, **3**, **5**, and **7** are summarized in Fig. 2. For clarity, the sites of strand cleavage were shown by the number of bases separated from the A-^{CNBP}U base pair with plus (toward 5′ side) and minus (toward 3′ side) signs.

The cleavage efficiency was obtained from relative band intensity versus the sum of total DNA band intensities. Remarkable strand cleavage was observed only for ODN **5** at the G of the position −4. Highly selective cleavage occurred at the 5′ G of the GG step, suggesting that the G cleavage proceeded via $G^{+\cdot}$. The efficiency of 5′ G oxidation of GG steps was sensitive to the position of the single G (shown in underlined bold face) proximal to the d^{CNBP}U. The cleavage efficiency decreased ca. one tenth when the single G moved one base pair away from position −1 (ODN **5**) to position −2 (ODN **7**). Only weak

Table 1 d^{CNBP}U- and GG-containing oligomers[a]

1:	5'-ATACTAACATT**GG**TTGATAACT-3'
2:	3'-TATGATTGTAACCAAC**X**ATTGA-5'
3:	5'-ATACTACATT**GG**TTGTATAACT-3'
4:	3'-TATGATGTAACCAACA**X**ATTGA-5'
5:	5'-ACTAATATTA**G**TT**GG**TTATGAT-3'
6:	3'-TGATTATAA**X**CAACCAATACTA-5'
7:	5'-ACTATATTATGTT**GG**TTATGAT-3'
8:	3'-TGATATAA**X**ACAACCAATACTA-5'
9:	5'-ATTTATA**G**TA**GG**TA**GG**TATTT-3'
10:	3'-TAAATA**X**CATCCATCCATAAA-5'
11:	5'-ATTTATA**G**TACCTA**GG**TATTT-3'
12:	3'-TAAATA**X**CAT**GG**ATCCATAAA-5'
13:	5'-ATTTATA**G**TGTGTA**GG**TATTT-3'
14:	3'-TAAATA**X**CACACATCCATAAA-5'
15:	5'-ATTTATA**G**TATATA**GG**TATTT-3'
16:	3'-TAAATA**X**CATATATCCATAAA-5'

[a] X represents d^{CNBP}U. The A-X base pair and the GG sites are shown in bold face and red.

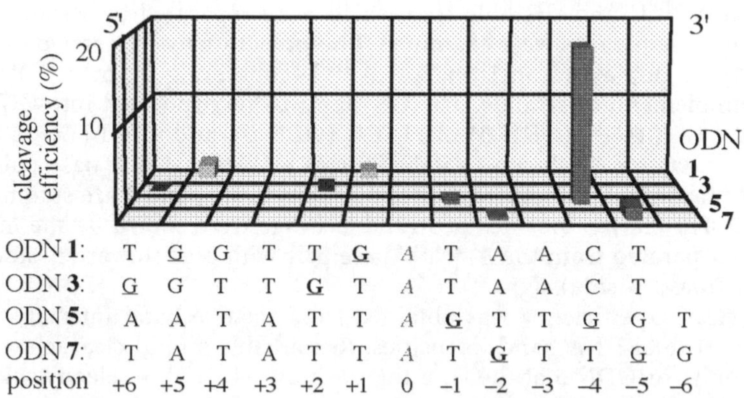

Fig. 2 Photocleavage of GG-containing oligomers complementary to the corresponding d^{CNBP}U-containing strands. d^{CNBP}U was located opposite to the A at position 0 (shown in red). Partial sequences of ODNs are shown and the sites of strand cleavage are underlined. Single Gs proximal to the d^{CNBP}U-A base pair are shown in bold face. Efficiencies at the major cleavage sites were 18.5% (at position –4, ODN 5), 1.9% (at position +5, ODN 1), and 1.5% (at position –5, ODN 7)

cleavage was observed for ODNs **1** and **3** containing the single G at positions +1 and +2, respectively. In a control experiment, photoirradiation of a duplex possessing d(ATTGG)/d(CCAA^{CNBP}U) sequence containing no such proximal single G resulted in no cleavage of the GG step, indicating that direct single electron transfer (SET) to the photoexcited d^{CNBP}U from Gs that are more than three bases away from the A-^{CNBP}U base pair is unfeasible. Since the intervening 5′TT3′ sequence between the single G and the remote GG sites was the same for ODNs **5** and **7**, the cleavage intensity at 5′ G of the GG step directly reflects the apparent efficiency of $G^{+\cdot}$ formation in both systems (position −1 and −2). These data showed that guanine radical cation (hole) is injected site selectively at the G in the sequence of C^{CNBP}U/AG by SET from G to photoexcited d^{CNBP}U in the complementary strand. The hole generated migrates to distal GG sites via hole transport (HT) and is eventually trapped by oxygen or water to give piperidine labile sites (Scheme 1).

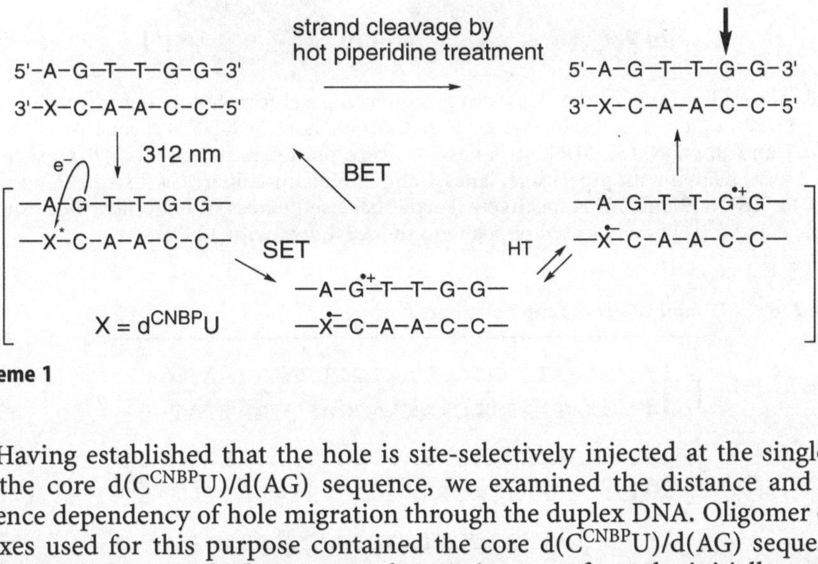

Scheme 1

Having established that the hole is site-selectively injected at the single G in the core d(C^{CNBP}U)/d(AG) sequence, we examined the distance and sequence dependency of hole migration through the duplex DNA. Oligomer duplexes used for this purpose contained the core d(C^{CNBP}U)/d(AG) sequence and a GG hole trap which was seven base pairs apart from the initially generated $G^{+\cdot}$ center. The intervening base sequences were designed so as to have two GG steps in the same strand (ODNs **9/10**) or one in an opposite strand (ODNs **11/12**). Other sequences consisted of two single Gs instead of one GG step (ODNs **13/14**) or only AT base pairs (ODNs **15/16**). Strand cleavage of ODN **9** occurred selectively at 5′ G of both proximal and distal GG steps $(GG)_p$ and $(GG)_d$, respectively (Fig. 3, lane 4). The relative band intensity $[GG]_d/[GG]_p$ was 0.84 in an average of seven separate experiments. Cleavage at $(GG)_d$ of duplex **11/12**, where $(GG)_p$ was in the opposite strand (ODN **12**), was roughly two-fold more efficient than that observed for duplex **9/10** possessing two GG steps and the $G^{+\cdot}$ center in the same strand (lane 5 vs lane 4).

In a separate experiment, we confirmed the cleavage at 5′ G of the GG step in opposite strand ODN **12**. The presence of G in the intervening sequence

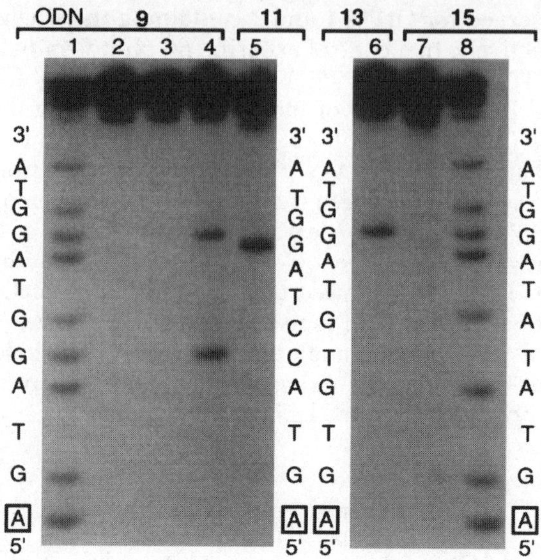

Fig. 3 Autoradiograms of the denaturing sequencing gel for photoreactions of duplexes **9/10**, **11/12**, **13/14**, and **15/16**. Lanes 1–4, ODN **9**; lane 5, ODN **11**; lane 6, ODN **13**; lanes 7 and 8, ODN **15**; ODNs in lanes 3–7 were photoirradiated; all ODNs except in lane 3 were heated with piperidine; lanes 1 and 8, Maxam-Gilbert G+A sequencing reactions for ODNs **9** and **15**, respectively. Partial base sequences of oligomers were shown on the side. $d^{CNBP}U$ was located opposite to the A (shown with a box)

Table 2 $d^{CNBP}U$- and GG-containing oligomers[a]

17:	5'-ATTAATGTTATT**GG**TTAATTAATA-3'
18:	3'-TAATTACAA**X**AACCAATTAATTAT-5'
19:	5'-ATTAATGTTAGT**GG**TTAATTAATA-3'
20:	3'-TAATTACAA**X**CACCAATTAATTAT-5'
21:	5'-ATTAATGTTAGTT**GG**TTAATTAAT-3'
22:	3'-TAATTACAA**X**CAACCAATTAATTA-5'
23:	5'-ATTAATGTTAGTTT**GG**TTAATTAA-3'
24:	3'-TAATTACAA**X**CAAACCAATTAATT-5'
25:	5'-ATTAATGTTAGTTTT**GG**TTAATTA-3'
26:	3'-TAATTACAA**X**CAAAACCAATTAAT-5'
27:	5'-ATTAATGTTAGTTTTT**GG**TTAATT-3'
28:	3'-TAATTACAA**X**CAAAAACCAATTAA-5'

[a] X represents $d^{CNBP}U$. The A-X base pair and the GG sites are shown in bold face and red.

was found to be essential for hole migration over 24 Å by comparing the cleavage at the GG step of ODN **13** (lane 6) with that of ODN **15** (lane 7). Therefore, the intervening 5′TGTGTA3′ sequence in ODN **13** is able to mediate hole migration, whereas the 5′TATATA3′ sequence in ODN **15** does not. Despite the formation of $G^{+\cdot}$ next to the A-^{CNBP}U base pair in the photoirradiation of ODNs **15/16**, only very weak cleavage was observed at this G. This suggests that the initially formed $G^{+\cdot}$ may be rapidly quenched by back electron transfer from d^{CNBP}U radical anion prior to the degradation of $G^{+\cdot}$ leading to piperidine sensitive product.

In a another experiment, we confirmed that the hole migration from the $G^{+\cdot}$ to the remote GG step through four intervening AT base pairs is extremely difficult (Table 2 and Fig. 4). These observations are consistent with

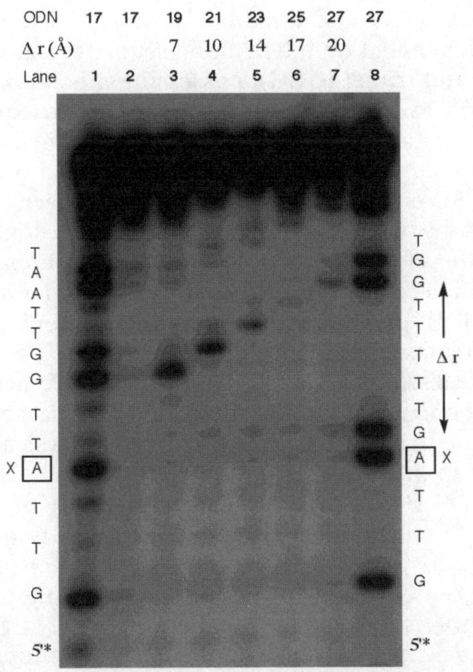

Fig. 4 An autoradiogram investigating the distance dependency of hole migration using ODNs **17–28**. The ^{32}P 5′-end labeled ODNs were hybridized to their complementary strands (5 μM, strand concentration) in 10 mM sodium cacodylate at pH 7.0. Hybridization was achieved by heating the sample at 90 °C for 5 min and slowly cooling to room temperature. The ^{32}P 5′-end labeled ODN duplexes were irradiated at 312 nm for 90 min with transilluminator at 0 °C. After piperidine treatment, the samples were suspended in denaturing loading buffer and electrophoresed through a denaturing 15% polyacrylamide/7 M urea gel. Lanes 1 and 8, Maxam-Gilbert sequencing reactions of ODNs **17** and **27**, respectively; lane 2, ODN **17** containing no initial hole site; lane 3, ODN **19**, lane 4, ODN **21**; lane 5, ODN **23**; lane 6, ODN **25**; lane 7, ODN **27**. Distances between the initial G+ center and the 5′ G of the GG step (Δr) were ca. 7, 10, 14, 17, and 20 Å for ODNs **19**, **21**, **23**, **25**, and **27**, respectively

results reported by Giese and co-workers that (i) long range CT in duplex DNA proceeds via a successive hole hopping process between G bases and (ii) the efficiency of each charge-transfer process rapidly decreases as the number of AT base pairs separating individual G bases increases.

3 Modulation of DNA-Mediated Hole Transport Efficiency Between Two G Triplets

3.1 The Effect of Intervening Sequence Between Two G Triplets [2]

The basis of the hopping mechanism is that a guanine radical cation ($G^{\cdot+}$) cannot oxidize adenine (A) due to the higher ionization potential (IP) of A compared with that of G, but can oxidize another G. We reported that i) the IP of G is highly dependent on the flanking sequences and ii) stacked Gs like a G doublet (GG) and triplet (GGG), possessing much lower IPs than that of isolated G, can serve as an effective hole trap. As a consequence, when a hole donor is a radical cation of G triplet $(GGG)^{\cdot+}$, an isolated G cannot be a hole acceptor due to the large free energy required for the process, but may act as a bridged base lowering the IP of the bridge between two G triplets. It is actually predicted by electron transfer theory that lowering the IP of a bridge increases the electronic coupling for the superexchange interaction between donor and acceptor [19–22]. In order to gain a deeper understanding of the effect of IP of bridged bases on HT efficiency, we examined HT between two G triplets separated by a bridge of TT*B*TT containing a bridged base (*B*) of A, 7-deazaA (^{z}A), G, or 7-deazaG (^{z}G). We herein report for the first time that the efficiency of DNA-mediated HT markedly increases with decreasing IP of the bridged base. Furthermore, ^{z}G was shown to be an extremely efficient trap in HT through the DNA π-stack.

The modulation of HT efficiency by changing the IP of a bridged base was investigated on a 29-mer duplex containing a probe sequence of 5′-A_1G_2T GTG_6 GGT T*B*T TG_{14}G G-3′ with a bridged base (*B*) of A (ODN **30**), ^{z}A (ODN **31**), G (ODN **33**), or ^{z}G (ODN **34**) (Table 3). Upon photoirradiation of the duplex, the hole was site-selectively generated at G_2 by a single electron transfer to a photoexcited cyanobenzophenone-substituted 2′-deoxyuridine ($d^{CNBP}U$) opposite A_1 in the complementary strand (ODNs **29** and **32**), and then irreversibly migrated to a proximal G triplet (G_6GG). A distal G triplet (G_{14}GG) is separated from a proximal G_6 triplet by five base pairs via TT*B*TT with a bridged base *B* in the middle.

The calculated IPs of GGG, A, ^{z}A, G, and ^{z}G at B3LYP/6–31G(d) were 4.17, 5.66, 5.25, 4.93, and 4.55 eV, respectively. Free energy changes (ΔG) for HT from $(GGG)^{\cdot+}$TT*B*TTGGG to GGGTT$B^{\cdot+}$TTGGG were estimated from the total energy change (ΔE) obtained from the following equation:

$$(GGG)^{\cdot+} + B \rightarrow G(GGG) + B^{\cdot+} + \Delta E$$

Table 3 d^{CNBP}U and GG-containing oligomers[a]

29 :	3'-M-**X**C ACAC CCAA TAAC CC-V-5'
30 :	5'-N-A**G$_2$**TGT**G$_6$GG**TT **A**TT**G$_{14}$GG**-W-3'
31 :	5'-N-A**G$_2$**TGT**G$_6$GG**TT^{z}**A**TT**G$_{14}$GG**-W-3'
32 :	3'-M-**X**C ACAC CCAA CAAC CC-V-5'
33 :	5'-N-A**G$_2$**TGT**G$_6$GG**TT **G**TT**G$_{14}$GG**-W-3'
34 :	5'-N-A**G$_2$**TGT**G$_6$GG**TT^{z}**G**TT**G$_{14}$GG**-W-3'

[a] X represents d^{CNBP}U. X, G_2, two G triplets, and bridged bases are shown in bold face. M = TAAATA, N = ATTTAT, V = AATAATA, W = TTATTAT

ΔG values (eV) were −1.76, −1.36, −0.98, and −0.56 for A, ^{z}A, G, and ^{z}G, respectively. A schematic illustration of the energy diagram for HT from $G_2^{\cdot+}$ to G_{14}GG is shown in Fig. 5.

5′-^{32}P-End-labeled oligomers **30**, **31**, **33**, and **34** were annealed with their complementary strands. Duplexes **29/30**, **29/31**, **32/33**, and **32/34** were photoirradiated at 312 nm for 1 h. G oxidation sites were determined by densitometric assay of the cleavage bands after hot piperidine treatment. As noted in our previous report [1], the G oxidation increases linearly with irradiation time. Normalized intensities of the cleavage bands are graphically shown in Fig. 6. G oxidation of ODN **30** having adenine as a bridged base occurred selectively at G_6 and G_7 in a proximal G_6 triplet. Cleavage intensities decreased in the order middle G (G_7)≫5′G (G_6), indicating a typical one electron oxidation at the TGGGT sequence [23]. Band intensity at G_{14}GG relative to that at G_6GG (I_{G14}/I_{G6}) was only 0.05, confirming previous observations that HT through five AT base pairs proceeds with extremely low efficiency. In sharp contrast, cleavage at G_{14} was observed for ODNs **31**

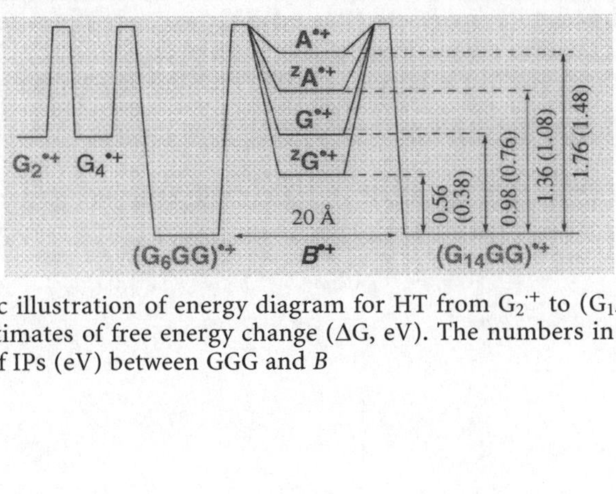

Fig. 5 Schematic illustration of energy diagram for HT from $G_2^{\cdot+}$ to $(G_{14}GG)^{\cdot+}$ via putative $B^{\cdot+}$ with estimates of free energy change (ΔG, eV). The numbers in parentheses are the difference of IPs (eV) between GGG and *B*

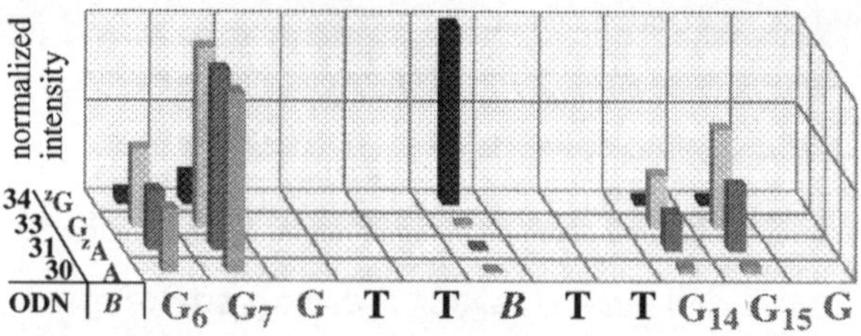

Fig. 6 Graphical illustration of normalized intensities of cleavage bands at G6, G7, B, G14, and G15 for oligomers **30**, **31**, **33**, and **34**. Data represents average of three data sets. Intensities are normalized so that the strongest cleavage is 1.00. I_{G14}/I_{G6} for ODNs **30**, **31**, and **33** were 0.05, 0.42, and 0.59, respectively

and **33** containing ^{z}A and G as a bridged base, respectively. I_{G14}/I_{G6} was 0.42 for ODN **31** (B=^{z}A) and increased to 0.59 for ODN **33** (B=G). The trajectory of HT dramatically changed when ^{z}G was incorporated into the bridge between two G triplets. Intensive cleavage of ODN **34** occurred selectively at ^{z}G but not at all at the G_{14} triplet. HPLC analysis of the nucleoside mixture of photoirradiated duplex d(GTCCACXATC)/d(GATAGT^{z}G GAC) after heating with piperidine showed a complete disappearance of ^{z}G, whereas more than 80% of X (d^{CNBP}U) was recovered unchanged together with almost quantitative recovery of A, C, G, and T. This indicates that ^{z}G is actually oxidized and decomposed to a piperidine labile site under the photoirradiation conditions. Furthermore, the cleavage at G_6 of ODN **34** was significantly suppressed compared with those of ODNs **30**, **31**, and **33**, indicating that ^{z}G not only terminates HT but also effectively drags a hole into its own site.

Our experiments described here clearly show that i) HT through a bridge of five AT base pairs proceeds with extremely low efficiency, ii) HT is effectively mediated when the bridge contains ^{z}A or G, iii) cleavage intensities at the proximal G triplet are much higher than those at the distal G triplet, iv) HT efficiency significantly increases by lowering the IP of the bridged base, and v) HT is terminated at the site of ^{z}G. Stronger cleavage at the proximal G_6 triplet than at the distal G_{14} triplet observed for ODNs **30**, **31**, and **33** indicates that the rate (k_{trap}) for trapping of $(GGG)^{\cdot+}$ with oxygen, eventually giving piperidine labile products (P_6 and P_{14}), exceeds the rate (k_{HT}) for HT (Scheme 2). Assuming a very weak directional preference of HT between two G triplets, the rate for HT relative to hole trapping for ODNs **30**, **31**, and **33**

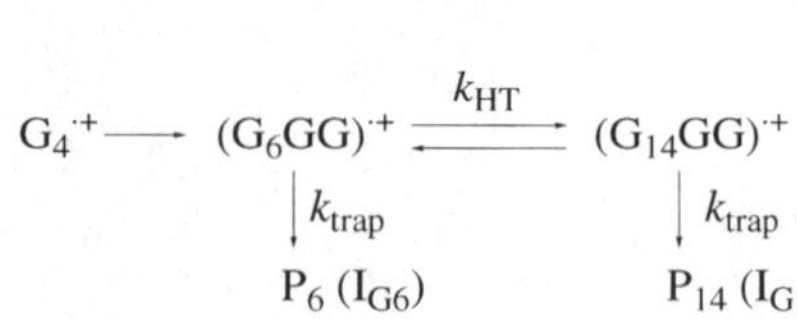

Scheme 2

would be estimated by the I_{G14}/I_{G6} value. Lowering the IP of a bridged base by 0.32 eV (from ^{z}A to G) increased I_{G14}/I_{G6} 1.4-fold. These results clearly show that HT efficiency is sensitively modulated by IPs of the bridged base, suggesting that HT between two G triplets with bridges of TT^{z}ATT and TTGTT proceeds via a superexchange mechanism. Lowering the IP of the bridged base increased the electronic coupling for the superexchange interaction between the two G triplets. In addition to the ΔG term, the reorganization energy (λ) of a bridged base radical cation would also affect electronic coupling. A large increase of the HT efficiency by replacing A with ^{z}A may suggest extra effects of decreased λ for $^{z}A^{\cdot+}$. Further lowering the IP at the bridged base, by replacing G with ^{z}G, resulted in HT from $(GGG)^{\cdot+}$ actually inducing oxidation of the bridged base, ^{z}G. Our calculations show that ^{z}G is not a better thermodynamic sink for HT than G triplet, suggesting that the selective cleavage of ODN **34** at ^{z}G is most likely due to a kinetic factor, for example the trapping rate of $^{z}G^{\cdot+}$ leading to a piperidine labile site would be significantly higher than that for $(GGG)^{\cdot+}$. These results show that the efficiency of HT through DNA π-stack is highly sequence dependent. Our results were also interpreted by a thermally induced hopping mechanism [24].

3.2
Suppression of DNA-Mediated Charge Transport by *Bam*H I Binding [5]

Guanine radical cation produced by one electron oxidation of DNA is not localized at the initially oxidized guanine site, but migrates over long distances through the π-stacked array of base pairs. This quite striking property of DNA has been thoroughly studied using oligomer duplexes containing various types of electron acceptors, showing DNA mediates charge transport in a highly sequence-dependent manner. As the number of observations of DNA-mediated charge transport using oligomer duplexes increases, fundamental questions regarding the likelihood of charge transport in genomic DNA, and its biological consequences, arise as the next issues to be studied. Chromosomal DNA in eukaryotes is stored in the nucleus as a form of chromatin with DNA bound to positively charged histone octamers. The X-ray structure of the nucleosome core particle shows that the structure of the DNA bound to histone deviates from the ideal superhelix geometry [25]. Distortion of the DNA structure changes the degree of base stacking. It has been shown that disruption of the DNA π-stack induced by protein binding decreases the charge transport efficiency [26, 27]. Besides the structural alteration, the protein binding has significant effects on the electronic state of DNA. In particular, the distribution of the electron density on nucleotide bases is modified by the hydrogen bonding of charged groups to nucleotide bases. However, the electronic effects of protein binding on the efficiency of charge transport through a DNA π-stack remain to be clarified. We reveal that binding of endonuclease *Bam*H I to its recognition sequence 5′-GGATCC-3′, involving hydrogen bonding of a positively charged guanidi-

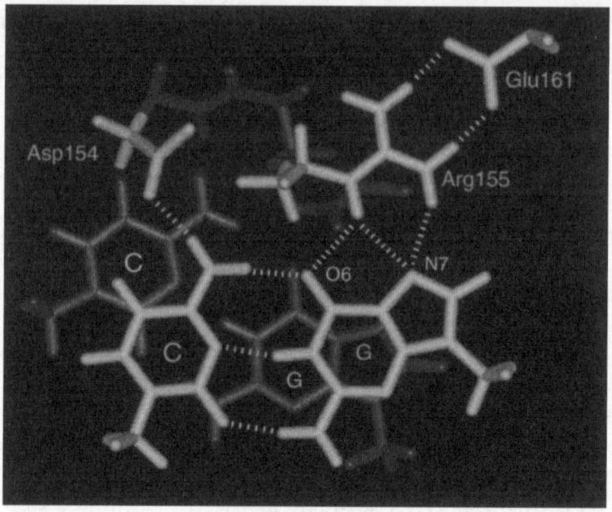

Fig. 7 Hydrogen bonding contacts between *Bam*H I and G-C base pairs in the recognition sequence of 5′-GGATCC-3′, taken from the X-ray structure of *Bam*H I-DNA complex reported by Luger et al. [25], and the effect of hydrogen bonding on the electrostatic potential. The protein-DNA interactions involving 5′ and 3′ side G-C base pairs are colored in yellow and blue, respectively. The N-7 and O-6 atoms of 5′ side G are directly hydrogen-bonded to a guanidium group of Arg155 that is also bound to Glu161. The O-6 of 3′ side G is bound to Asn116. For clarity, a hydrogen bonding network involving a 5′ side G-C base pair is shown with a dotted line

um group to guanine, effectively suppresses the oxidation of the sequence and the charge transport through the binding site.

*Bam*H I is a restriction endonuclease that binds as a dimer to the palindromic sequence 5′-GGATCC-3′, and hydrolyses the phosphodiester linkage between the two guanines in the presence of Mg^{2+} [28]. The X-ray structure of the *Bam*H I-DNA complex shows that direct hydrogen bonding involved in the protein-DNA contacts are condensed in a major-groove face of two G-C base pairs (Fig. 7) [25]. The protein-bound DNA retains a standard B-form-like conformation without significant bends and distortions of the base stack. In contrast, the electronic state of the protein-bound DNA seems significantly different from the free-state. Direct hydrogen bonding of a positively charged guanidium group of Arg155 of *Bam*H I to both N-7 and O-6 of the 5′ guanine of the 5′-GGATCC-3′ binding motif should make the electron density of the guanine in the complex lower than in the free-state DNA. Electrostatic potentials of free guanine and guanine-Arg155-Glu161 triad in the *Bam*H I-DNA complex were calculated at the level of B3LYP/6–31G(d). Geometries of the molecules were obtained from the coordinates of the X-ray structure of the *Bam*H I-DNA complex deposited in the protein data bank, and used for the calculations without any change. The electrostatic potentials were mapped on the surface of a total electron density (0.002 electrons/au^3) of each molecule. The electrostatic potential mapped on the sur-

Table 4 Sequences of oligomer duplexes containing a *Bam*H I recognition sequence (GGATCC)

			8	11	13	16 17	21				
35:	5'-ATT	TAT	A**G**T	AGT	GAT	**GGA**	**TCC**	ATA	TTA	T-3'	
36:	3'-TAA	ATA	**X**CA	TCA	CTA	**CCT**	**AGG**	TAT	AAT	A-5'	
							9 8				
			8	11	13	16	21	24	25 26		
37:	5'-ATT	TAT	A**G**T	AGT	GAT	**GGA**	**TCC**	AT**G**	**GG**A	TTA	T-3'
38:	3'-TAA	ATA	**X**CA	TCA	CTA	**CCT**	**AGG**	TAC	CCT	AAT	A-5'

[a] X represents $d^{CNBP}U$.

face of total electron density in fact shows that guanine in the complex is much more electron deficient than free guanine. Negative electrostatic potential that appeared at the region of N-7 and O-6 of the free guanine disappeared in the guanine-guanidium-carboxylate triad. We anticipated that such guanines with the decreased electron density become less easily oxidized due to the increase of the ionization potential, when compared with normal guanine, and no longer function as a stepping stone in the charge transport via guanine hopping.

To address the electronic effects of protein binding on the efficiency of charge transport through the DNA π-stack, we used oligomer duplexes **35/36** and **37/38**, which both include a *Bam*H I binding site in the middle of the sequence (Table 4). Oligomers **36** and **38** contain *p*-cyanobenzophenone substituted uridine ($d^{CNBP}U$) as an electron-accepting nucleotide base that initiates one electron oxidation of a neighbouring guanine (G_8 in ODNs **35** and **37**) upon irradiation at 312 nm. Stacked guanine sites of GG and GGG have lower oxidation potential compared to a single guanine, and are frequently used as internal hole traps. Therefore, a radical cation produced at G_8 can migrate through the π-stack down to $G_{16}G_{17}$ in ODN **35** and to G_8G_9 in ODN **36**, which both directly contact a guanidium group in Arg155 of the *Bam*H I-DNA complex.

We first examined the binding of *Bam*H I to duplex **35/36**, and its effect on the oxidation of $G_{16}G_{17}$ at the binding site. Optimum conditions for the complete complex formation between *Bam*H I and duplex **35/36** were determined by electrophoretic mobility shift assay. The *Bam*H I-DNA complex increased as the concentration of the protein increased. At a *Bam*H I concentration of 1.2 U/μL, duplex **35/36** (<2 nM) was completely transformed into the complex. These conditions (*Bam*H I, 1.2 U/μL; duplex, <2 nM) that ensure a complete complex formation were used for all of the *Bam*H I binding experiments described below. The effect of *Bam*H I binding on the oxidation of the 5′ GG in the recognition sequence was examined as a function of *Bam*H I concentration. In the absence of the protein, distinct and strong cleavage was observed at G_{16}, with additional weak cleavages at G_{13} and G_{17}.

It is well established in one electron oxidation of DNA that: i) a GG site is more strongly oxidized than a GA site, and ii) oxidation of a GG site is selective for the 5′ guanine. These observations ensure that G_{16} cleavage of ODN **35** proceeds by one electron oxidation. By increasing the concentration of *Bam*H I, the band intensity at G_{16} decreased and eventually faded away in the presence of 1.2 U/μL of the protein. Intensity of the cleavage band at G_{16} increased as the fraction of *Bam*H I-**35/36** complex increased. Complete disappearance of the G_{16} band demonstrated that $G_{16}G_{17}$ was insulated from one electron oxidation upon complex formation. In marked contrast to G_{16}, the band intensity at G_{13} in a GA site increased with protein concentration. These opposing results regarding the efficiency of the oxidation of G_{16} and G_{13} suggest that the $G_{13}A$ site of duplex **35/36** becomes the most easily oxidized site in the *Bam*H I bound duplex, because of the insulation of $G_{16}G_{17}$ by the guanidium group.

Since *Bam*H I binds as a dimer to the palindromic sequence of 5′-GGATCC-3′, two GG sites in the sequence should be equally insulated from one electron oxidation. In the absence of the protein, both $G_{16}G_{17}$ in ODN **35** (Fig. 8a, lane 2) and G_8G_9 in ODN 2 (Fig. 8b, lane 2) showed similar oxidization patterns under the irradiation conditions. In contrast, cleavage bands at both GG sites completely disappeared in the presence of *Bam*H I (1.2 U/μL) (lane 3 in Fig. 8a,b). Simultaneous suppression of oxidation at both GG sites shows that insulation of both GG sites from one electron oxidation is due to the binding of *Bam*H I to the recognition sequence.

As we had established that *Bam*H I binding suppressed one electron oxidation of GG in the recognition sequence, we then investigated charge transport through the site of *Bam*H I binding. Since a hole injecting G_8 and a hole trapping $G_{24}G_{25}G_{26}$ in duplex **37/38** are located on opposite sides of the *Bam*H I binding site, hole migration from G_8 to the G-triplet must proceed through the site of *Bam*H I binding. In the absence of the protein, the guanine radical cation produced at G_8 in the duplex **37/38** migrates down to $G_{24}G_{25}G_{26}$ through the $G_{16}G_{17}ATCC$ sequence. Strong cleavage bands are observed at G_{24} and G_{25} in the G-triplet, in addition to the cleavage at G_{16} of the GG site. Only a faint band was observed at G_{13} in the GA site. In the presence of *Bam*H I, band intensity at the G-triplet decreased with the concomitant disappearance of the band at G_{16}. Densitometric analysis showed that oxidation of the G triplet in the protein bound duplex was suppressed more than 3.7-fold (Fig. 9). However, accurate analysis was not feasible due to very weak cleavages at the G triplet in the complex. In spite of the presence of the hole trapping G-triplet, the predominant site for one electron oxidation in the protein bound duplex is the $G_{13}A$ site, suggesting a considerable decrease in the efficiency of the charge transport from G_{13} to G_{24}. These results clearly show that *Bam*H I binding to DNA not only suppressed the one electron oxidation of GG in the recognition sequence, but also lowered the charge transport through the site of the protein binding. This remarkable effect of *Bam*H I binding on DNA mediated charge transport is most ascribable to electronic insulation of the guanine in the binding site by direct con-

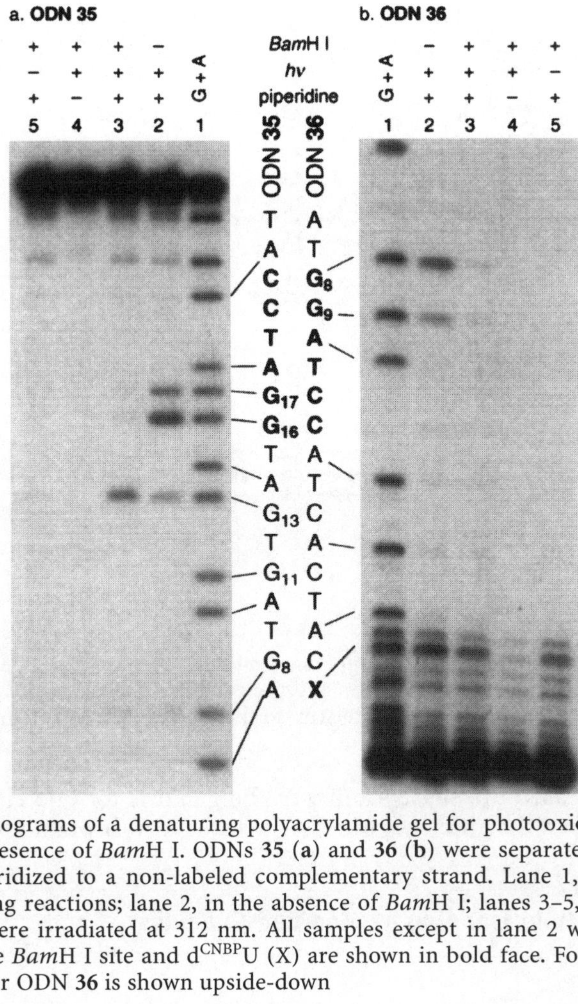

Fig. 8 Autoradiograms of a denaturing polyacrylamide gel for photooxidation of duplex **35**/**36** in the presence of *Bam*H I. ODNs **35** (**a**) and **36** (**b**) were separately 5′-^{32}P-end labeled and hybridized to a non-labeled complementary strand. Lane 1, Maxam-Gilbert G+A sequencing reactions; lane 2, in the absence of *Bam*H I; lanes 3–5, *Bam*H I. ODNs in lanes 2–4 were irradiated at 312 nm. All samples except in lane 2 were heated with piperidine. The *Bam*H I site and d^{CNBP}U (X) are shown in bold face. For clarity, the autoradiogram for ODN **36** is shown upside-down

tact of the positively charged guanidium group of the protein via hydrogen bonding.

The electrostatic contacts between positively charged amino acid residues and negatively charged DNA are extremely important for protein-DNA interactions. Non-specific electrostatic contacts of proteins are mostly to the phosphate anion of DNA backbone, but sequence-specific protein bindings involve direct or water-mediated hydrogen bonding of charged groups to nucleotide bases. Contact via hydrogen bonding of a guanidium group in arginine to guanine is one of the most commonly observed protein-DNA interactions and is indispensable for the sequence recognition. Our results described here show that direct contact of a guanidium group to guanine dra-

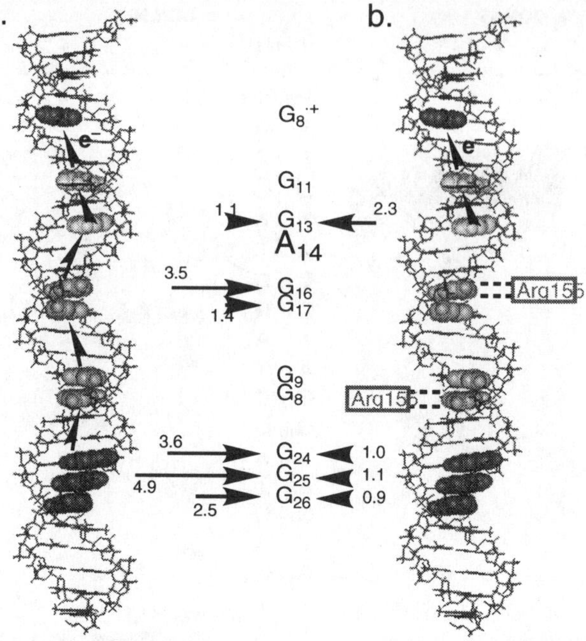

Fig. 9 Illustrations of the charge transport in duplex **37/38**, **a** in the absence and **b** in the presence of *Bam*H I. Horizontal arrows and the numbers shown on the site of guanine oxidation indicate the band intensity relative to that of G24 in the protein bound duplex

matically decreased the susceptibility of the guanine for one electron oxidation, and lowered the efficiency for charge transport through the guanine.

4 Site-Selective Hole Trapping by Modified Guanines

4.1 Termination of DNA Mediated Hole Transport at N^2-Cyclopropyldeoxyguanosine [4]

Due to lower ionization potentials of GG and GGG than single G, these stacked G sites function as a thermodynamic sink of holes eventually producing piperidine labile sites. In principle, hole migration from hole donor to acceptor competes with hole trapping by water and/or oxygen. Therefore, overall efficiency of hole transfer is primarily determined by the rates of hole migration and hole trapping. When the rate of hole trapping is much slower than hole migration rate, equilibration of hole between donor and acceptor can be achieved [29]. While the rate of hole migration can be attenuated by changing the potential energy gap between hole donor and acceptor [30], the modulation of hole transport efficiency by changing the rate of hole trapping has never been demonstrated. We demonstrated a novel hole-trapping nucleoside N^2-cyclopropyl-2′-deoxyguanosine (d^{CP}G), which possesses a cyclopropyl group on N^2 as a radical-trapping device. One electron oxidation of $d^{CP}G$ by photoexcited riboflavine induces homolytic cyclopropane ring opening, as evidenced by the formation of N^2-(3-hydroxypropanoyl)dG. With the use of $d^{CP}G$-containing duplex, we have demonstrated that $d^{CP}G$ efficiently terminates DNA-mediated hole transport at its own site.

It is known that radical cations of cyclopropylamine [31–33] and *N*-alkyl- and *N*-arylcyclopropylamines [34–36] rapidly undergo homolytic cyclopropane ring opening to produce *β*-iminium carbon radicals. The rate of homolytic ring opening of the cyclopropylamine radical cation is believed to be larger than that of the corresponding ring opening of neutral *N*-alkylcyclopropylaminyl radical (7.2×10^{11} s^{-1}) [37]. Molecular orbital calculations of the radical cation of N^2-cyclopropyl-9-methylguanine at B3LYP/6–31G(d) showed that i) the bond length of C-1"–C-2" in cyclopropane increased by 0.01 Å from its neutral state and ii) spin density is actually located on C-2" (Fig. 10). While the magnitude of the rate is unknown, the cyclopropane ring opening of the $d^{CP}G$ radical cation is expected to be rapid. We first examined one electron oxidation of $d^{CP}G$ with photoexcited riboflavin in aqueous solution. $d^{CP}G$ was rapidly consumed by photoirradiation at 366 nm in the presence of riboflavin, producing two major products after subsequent incubation of the photoirradiated mixture. These products were identified as dG and N^2-(3-hydroxypropanoyl)dG by ^{1}H NMR and high resolution FABMS (Scheme 3). Comparison of ^{1}H NMR of N^2-(3-hydroxypropanoyl)dG with

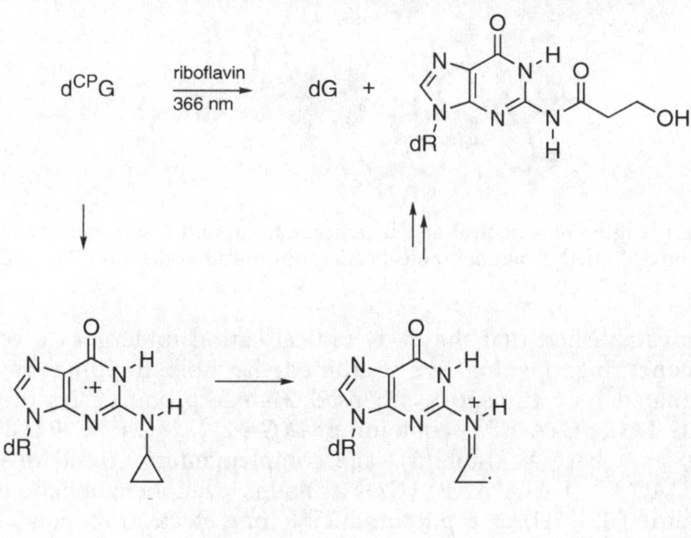

Scheme 3

the authentic sample unambiguously confirmed the structure. Formation of dG by incubating the photoirradiated mixture suggested that one electron oxidation of $d^{CP}G$ activated the transformation of cyclopropyl group at N^2 to a group being highly susceptible to hydrolysis.

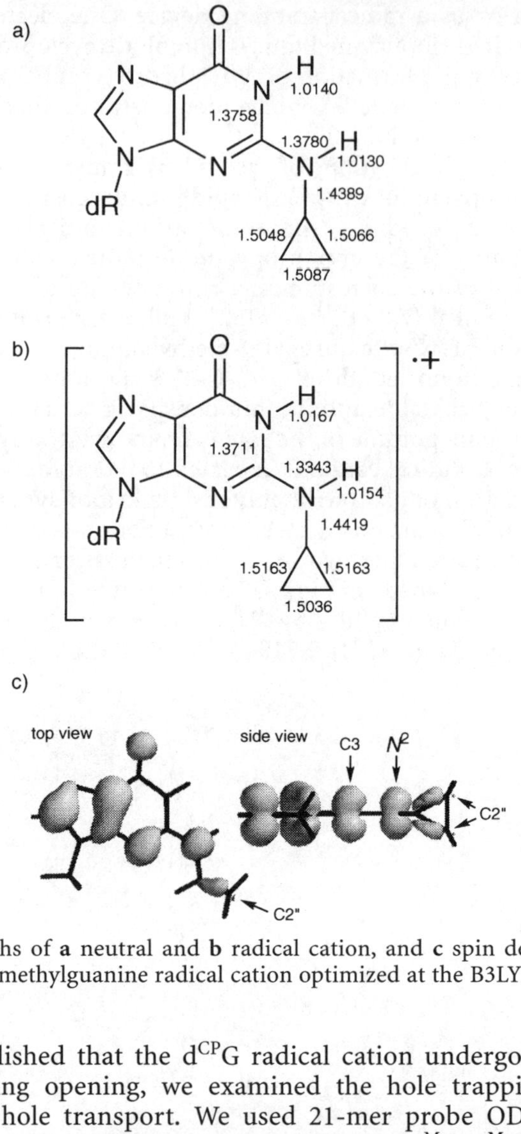

Fig. 10 Bond lengths of **a** neutral and **b** radical cation, and **c** spin density mapping of N^2-cyclopropy-N^9-methylguanine radical cation optimized at the B3LYP/6–31G(d) level

Having established that the d^{CP}G radical cation undergoes a very rapid cyclopropane ring opening, we examined the hole trapping by d^{CP}G in DNA mediated hole transport. We used 21-mer probe ODNs d(ATT TAT AG$_8$T **X**TG TA**G$_{15}$ G**TA TTT) containing G (**G-P21**), ^{Me}G ($^{\mathbf{Me}}$**G-P21**), and ^{CP}G ($^{\mathbf{CP}}$**G-P21**) as a base **X** (Table 5). The complementary strand d(AAA TAC CTA CAC AC^{CNBP}U ATA AAT) (**C21**) contains cyanobenzophenone substituted uridine (d^{CNBP}U) as a photoinducible one electron oxidant. The guanine radical cation was site-selectively produced at G$_8$ in duplexes **G-P21/C21**, $^{\mathbf{Me}}$**G-P21/C21**, and $^{\mathbf{CP}}$**G-P21/C21** by single electron transfer to photoexcited d^{CNBP}U [1]. The hole is able to migrate up to G$_{15}$G through a bridge of d(TXTGTA)/d(TACACA). Probe ODNs $^{\mathbf{Me}}$**G-P21** and $^{\mathbf{CP}}$**G-P21** were obtained from 2-fluoroinosine containing 21-mer according to the reported method

Table 5 $d^{CNBP}U$- and GG-containing oligomers[a]

G-P21 :	5'-ATT TAT AG$_8$T GTG TAG$_{15}$ GTA TTT-3'
^{Me}G-P21 :	5'-ATT TAT AG$_8$T ^{Me}GTG TAG$_{15}$ GTA TTT-3'
^{CP}G-P21 :	5'-ATT TAT AG$_8$T ^{CP}GTG TAG$_{15}$ GTA TTT-3'
C21 :	5'-AAA TAC CTA CAC ACX ATA AAT-3'

[a] X represents $d^{CNBP}U$.

[38, 39]. Photoirradiation of duplex **G-P21/C21** at 366 nm for 15 min followed by subsequent piperidine heating produced a distinct cleavage band at G_{15} selectively. This indicates an efficient hole migration from G_8 to G_{15}. In marked contrast, G_{15} oxidation was significantly suppressed in **^{CP}G-P21/C21** duplex, where $d^{CP}G$ was incorporated into the bridge of d(T^{CP}GTGTA)/d(TACACA). Normalized band intensities at G_{15} relative to intact bands as determined by densitometry were 1.00 for **G-P21** (standard), 0.53 for **^{Me}G-P21**, and 0.09 for **^{CP}G-P21**. While incorporation of methyl group at N^2 of dG resulted in a modest reduction of G_{15} oxidation via hole transport from G_8 to G_{15}, the incorporation of the cyclopropyl group was dramatically effective for the suppression of hole transport.

The melting temperature is only lower by 1.3 °C for ^{CP}G-containing 10-mer duplex d(GTC CAC TAT C)/d(GAT A^{CP}GT GGA C) than for the corresponding G-containing 10-mer duplex (50 μM base concentration, 100 mM NaCl). The CD spectrum of the ^{CP}G-containing duplex shows a typical B-form structure. These observations imply that the disruption of the π-stack by incorporating $d^{CP}G$ is not the reason for the suppression of hole transport. HPLC analysis of a nucleoside mixture obtained by enzymatic digestion of photoirradiated **^{CP}G-P21/C21** clearly showed that $d^{CP}G$ was completely consumed under the photoirradiation conditions used for the PAGE experiments, although other nucleosides including $d^{CNBP}U$ remain largely unchanged. These results suggest that suppression of G_{18} oxidation in **^{CP}G-P21/C21** is most likely due to the selective destruction of $d^{CP}G$. Calculated ionization potential at B3LYP/6–31G(d) level is lower by only 0.13 eV for the N^9-methyl-^{CP}G/N^9-methyl-C base pair than for the normal G/C base pair, and higher by 0.40 eV than for GG/CC doublet. Oxidation potential of $d^{CP}G$ measured in water containing 0.1 M $LiClO_4$ was 0.93 V (vs. SCE, cf. G 1.07 V). Therefore, it is unlikely that $d^{CP}G$ functions as a thermodynamic sink in hole migration. On the basis of these observations, we propose that $d^{CP}G$ effectively terminates hole transport through the DNA π-stack by a rapid and irreversible cyclopropane ring opening of its radical cation.

The pseudo first-order rate of trapping of $G^{\cdot+}$ by water was estimated to be $6{\times}10^4$ s^{-1}, which is significantly smaller than the rate for the hole migration between two single Gs separated by two AT base pairs ($2.5{\times}10^6$ s^{-1}) [29]. Discovery of a powerful hole trapping nucleobase possessing a faster trapping rate than $2.5{\times}10^6$ s^{-1} would change a current view of hole transport. Studies described here clearly demonstrate that d^{CP}G can serve as a useful probe for gaining deeper insight into kinetic aspects of hole transport through the DNA π-stack. We have recently reported N^6-cyclopropyl-2′-deoxyadenosine as a new hole-trapping nucleoside [40].

4.2
Hole annihilation at *N*^{2}-Phenyldeoxyguanosine [6]

In a previous chapter, we discussed the termination of charge transport by increasing the hole trapping rate at the site of *N*^{2}-cyclopropyldeoxyguanosine (d^{CP}G), in which the cyclopropane ring opening functioned as a radical trapping device. These results imply that substitution of the exocyclic amino group of deoxyguanosine (dG) with functional groups could be an intriguing tool to modulate the reactivity of dG toward one-electron oxidation. We here show that incorporation of *N*^{2}-phenyldeoxyguanosine (d^{Ph}G) into duplex DNA dramatically suppresses oxidative decomposition, not only at d^{Ph}G, but more importantly, at dGs remote from the modified guanine.

The DNA oligomers containing d^{Ph}G and complementary strands used for the studies were listed in Table 6. All d^{Ph}G-containing ODNs were synthesized from corresponding ODNs containing 2-fluoroinosine by substitution of fluorine with aniline [38, 39]. d^{Ph}G was incorporated in probe ODN **^{Ph}GG1** by replacing G_{15} of the $G_{15}G$ doublet in the 21-mer ODN **GG1**. Complementary ODN **^{CNBP}U1** contains cyanobenzophenone substituted uridine (d^{CNBP}U) as a photoinducible one-electron oxidant [1], whereas ODN **T1** contained thymidine in place of d^{CNBP}U. In duplexes of **GG1/^{CNBP}U1** and **^{Ph}GG1/^{CNBP}U1**, the guanine radical cation was site-selectively produced at G_8 by one-electron transfer to photoexcited d^{CNBP}U. ODN **GG5** contained five GG doublets with identical neighboring sequences. The ^{Ph}GG site was

Table 6 Oligomers used for the studies[a]

GG1:	5'-**X** A**G**$_8$TGTGTA **G**$_{15}$**G V**-3'
^{Ph}GG1:	5'-**X** A**G**$_8$TGTGTA^{Ph}**G**$_{15}$**G V**-3'
T1:	3'-**Y** TC ACACAT C C **W**-5'
^{CNBP}U1:	3'-**Y** ^{CNBP}UC ACACAT C C **W**-5'
GG5:	5'-**X** A **G**$_8$**G**TA**G**$_{12}$**G**TA **G**$_{16}$**G**TA**G**$_{20}$**G**TA**G**$_{24}$**G V**-3'
^{Ph}GG5(8):	5'-**X** A^{Ph}**G**$_8$**G**TA**G**$_{12}$**G**TA **G**$_{16}$**G**TA**G**$_{20}$**G**TA**G**$_{24}$**G V**-3'
^{Ph}GG5(16):	5'-**X** A **G**$_8$**G**TA**G**$_{12}$**G**TA^{Ph}**G**$_{16}$**G**TA**G**$_{20}$**G**TA**G**$_{24}$**G V**-3'
c-GG5:	3'-**Y** T C CATC CAT C CATC CATC C **W**-5'
GG4:	5'-**X** A **G**$_8$**G**TA**G**$_{12}$**G**TA TA TA**G**$_{20}$**G**TA**G**$_{24}$**G V**-3'
^{Ph}GG4(8):	5'-**X** A^{Ph}**G**$_8$**G**TA**G**$_{12}$**G**TA TA TA**G**$_{20}$**G**TA**G**$_{24}$**G V**-3'
c-GG4:	3'-**Y** T C CATC CAT AT ATC CATC C **W**-5'

[a] **X** = ATTTAT, **Y** = TAAATA, **V** = TATTT, **W** = ATAAA

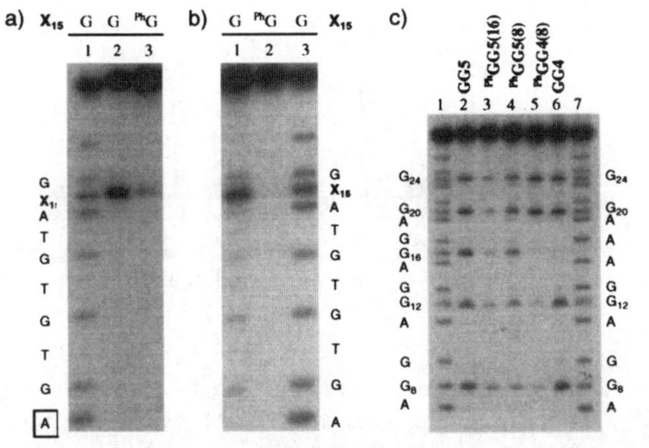

Fig. 11 Autoradiograms of denaturing sequencing gels for photoreactions of d^{Ph}G-containing oligomer duplexes. Photoirradiated ODNs were heated with piperidine and electrophoresed through a denaturing 15% polyacrulamide/7 M urea gel. Partial base sequences of ODNs were shown on the side. d^{CNBP}U was located opposite A (shown with a box). **a** Duplexes were irradiated at 312 nm for 60 min. Lane 1, Maxam-Gilbert A+G sequencing reactions of **GG1**; lane 2, ^{CNBP}**U1/GG1**; lane 3, ^{CNBP}**U1/**^{Ph}**GG1**. **b** Duplexes were irradiated at 366 nm in the presence of riboflavin. Lane 1, **T1/GG1**; lane 2, **T1/**^{Ph}**GG1**; lane 3, A+G reactions of **GG1**. **c** Duplexes were irradiated at 366 nm in the presence of riboflavin. Lane 1, A+G reactions of **GG5**; lane 2, **GG5/c-GG5**; lane 3, ^{Ph}**GG5(16)/c-GG5**; lane 4, ^{Ph}**GG5(8)/c-GG5**; lane 5, ^{Ph}**GG4(8)/c-GG4**; lane 6, **GG4/c-GG4**; lane 7, A+G reactions of **GG4**

embedded in ^{Ph}**GG5(8)** and ^{Ph}**GG5(16)** by replacing G_8 and G_{16} of **GG5** respectively with d^{Ph}G. ODNs **GG4** and ^{Ph}**GG4(8)** lacked the G_{16}G site of **GG5** and ^{Ph}**GG5(8)**, respectively, by replacing them with the T_{16}A sequence. **c-GG5** and **c-GG4** are complementary strands to **GG5** and **GG4**, respectively. The oxidation potential of d^{Ph}G measured by cyclic voltammetry in DMF containing 0.1 M $LiClO_4$ was 0.70 V (vs Ag/Ag^+, cf. G 0.67 V). The melting temperature of the d^{Ph}G-containing 10-mer duplex d(GAT AGT ^{Ph}GGA C)/d(GTC CAC TAT C) was 3.7 °C higher than the corresponding normal G-containing duplex (50 μM base concentration, 100 mM NaCl). CD spectra of the duplex showed a typical B-form structure.

Photoirradiation of duplex **GG1/**^{CNBP}**U1** at 312 nm for 60 min and subsequent piperidine treatment (90 °C, 20 min) produced a distinct cleavage band at the 5′ side G of the G_{15}G doublet (Fig. 11a, lane 2). In marked contrast, the cleavage at $^{Ph}G_{15}$ in duplex ^{Ph}**GG1/**^{CNBP}**U1** (lane 3) was considerably weaker than the cleavage at G_{15} in **GG1/**^{CNBP}**U1** (lane 2). The band intensity of G_{15} relative to intact full length **GG1** was 0.31, whereas the relative band intensity of $^{Ph}G_{15}$ of ^{Ph}**GG1** was only 0.05. Suppression of the decomposition of d^{Ph}G by one electron oxidation was further confirmed by riboflavin-sensitized oxidation of duplex ^{Ph}**GG1/T1**. While strong cleavage occurred at G_{15} of **GG1/T1** (Fig. 11b, lane 1), only a faint band was observed at $^{Ph}G_{15}$ of

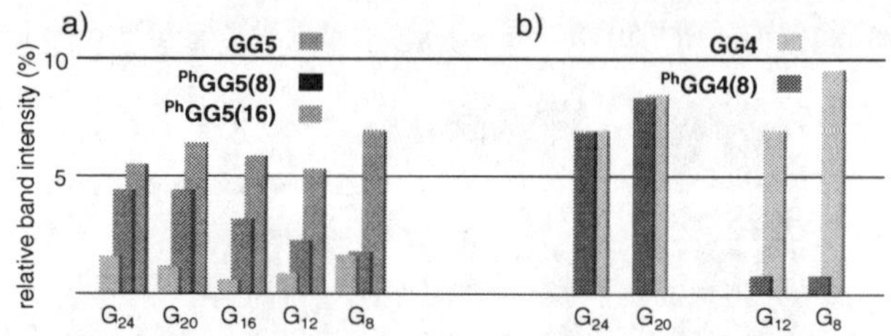

Fig. 12 Band intensities (%) of GG sites relative to the intact full length bands obtained for the photoirradiated duplexes of **a** lanes 2, 3, and 4, and **b** lanes 5 and 6 shown in Fig. 11c

^{Ph}GG1 (lane 2). We separately confirmed that strand cleavage of single stranded d^{Ph}G-containing oligomer d(ATT TAT AGT AGT AGT A^{Ph}GT ATT T) actually occurred at d^{Ph}G by the riboflavin-sensitized oxidation and subsequent piperidine treatment. The efficiency of the cleavage at d^{Ph}G was comparable to that at dG. Nucleoside analysis of d^{Ph}G-containing DNA by HPLC showed that 68% of d^{Ph}G remained intact in the duplex after 1 h irradiation in the presence of riboflavin, whereas only 36% of d^{Ph}G remained intact in the single strand.

Oxidative decomposition of the d^{Ph}G-containing duplex was suppressed not only at d^{Ph}G but also remarkably at the GG sites that were distant from d^{Ph}G (Fig. 11c). Band intensities at GG sites of lanes 1–5 in Fig. 11c are shown in Fig. 12. While strand cleavage of photoirradiated **GG5** in the presence of riboflavin occurred at all GG sites with comparable efficiency, the cleavage of GG sites of **^{Ph}GG5(8)** was suppressed at G_{12}G and G_{16}G in addition to $^{Ph}G_8$G (Fig. 12a). In contrast to G_{12}G, cleavage at G_{24}G was only weakly suppressed compared with the cleavage in **GG5**, showing that the efficiency of suppression of strand cleavage decreased with an increase in distance from d^{Ph}G. Distance dependency of cleavage suppression was clearly shown in the oxidation of **^{Ph}GG5(16)** that contained d^{Ph}G in the middle of five GG sites. The efficiency of the cleavage was considerably reduced at all four GG sites in addition to the ^{Ph}GG site. Significant insights into the mechanism of cleavage suppression by d^{Ph}G were obtained by the riboflavin-sensitized oxidation of **GG4** and **^{Ph}GG4(8)**. Strand cleavage of **GG4** was observed at all four GG sites, whereas the cleavage of **^{Ph}GG4(8)** was strongly suppressed at G_8G and G_{12}G sites, but not at all at G_{20}G and G_{24}G sites (Fig. 12b). The G_{12}G and G_{20}G sites in **^{Ph}GG4(8)/c-GG4** were separated by six intervening A-T base pairs, and the rate of hole transport between two sites was expected to be much smaller than that between G_8 and G_{12} (for instance 2.5×10^6 s^{-1} for $G^{\cdot+}$TTG→GTT$G^{\cdot+}$) and the estimated rate of hole trapping with water at the G site (6×10^4 s^{-1}).

Therefore, it is apparent that the efficiency of suppression of the oxidative decomposition at the given G sites in d^{Ph}G-containing duplex increased upon increasing the rate of hole transfer to d^{Ph}G. These remarkable observations could be rationalized by assuming an annihilation process of the d^{Ph}G radical cation that was prevented from decomposing, leading to the formation of a piperidine-labile site (Scheme 4). The rate of annihilation of the

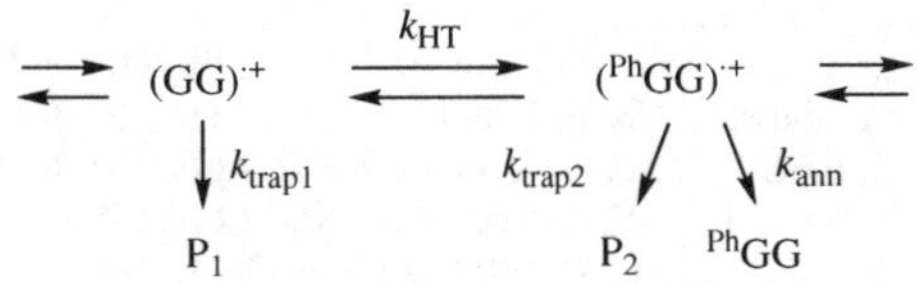

P_1, P_2 denote piperidine labile products.

Scheme 4

$d^{Ph}G$ radical cation (k_{ann}) should be much faster than the hole trapping rate at GG (k_{trap1}) and ^{Ph}GG (k_{trap2}) to suppress the decomposition at these sites. When the hole transfer to $d^{Ph}G$ overrode the hole trapping at the given G site ($k_{HT} >> k_{trap1}$), the hole was selectively depleted at $d^{Ph}G$ by the annihilation process. Molecular modeling simulation showed that the phenyl group of $d^{Ph}G$ in duplex DNA was located just in the middle of the minor groove, suggesting a considerable increase of solvent accessible surface upon phenyl substitution. The calculated spin density of the $d^{Ph}G$ radical cation was delocalized on the both purine and phenyl rings. Since suppression of $d^{Ph}G$ decomposition was specific for duplex DNA, the annihilation process most likely occurred in the vicinity of the phenyl ring in the minor groove. One plausible mechanism of the putative annihilation process may involve a back-electron transfer from superoxide radical anion to a $d^{Ph}G$ radical cation. In the case of riboflavin sensitization, the riboflavin radical anion is also conceivable as an electron donor.

Our studies showed that i) substitution of an exocyclic amino group of dG is effective in modulating the chemical properties of dG toward one-electron oxidation, and ii) decomposition of the guanine radical cation was effectively suppressed near $d^{Ph}G$. These results indicate that $d^{Ph}G$ is a prototype of nucleosides functioning as an intrinsic antioxidant of duplex DNA toward one-electron oxidation.

5 Conclusion

Incorporation of the modified nucleotide bases enables us to modulate the DNA properties that are extremely important to the charge transport efficiency. The data obtained by these experiments provides a much deeper insight and understandingof the mechanism of DNA mediated charge transport.

Acknowledgement The authors extend their deep appreciation to Dr. Chikara Dohno for his efforts, carrying out all of the studies described here through outstandingly well-thought-out experiments.

References

1. Nakatani K, Dohno C, Saito I (1999) J Am Chem Soc 121:10854–10855
2. Nakatani K, Dohno C, Saito I (2000) J Am Chem Soc 122:5893–5894
3. Nakatani K, Dohno C, Saito I (2000) Tetrahedron Lett 41:10041–10045
4. Nakatani K, Dohno C, Saito I (2001) J Am Chem Soc 123:9681–9682
5. Nakatani K, Dohno C, Ogawa A, Saito I (2002) Chem Biol 9:361–366
6. Nakatani K, Dohno C, Saito I (2002) J Am Chem Soc 124:6802–6803
7. Dohno C, Nakatani K, Satio I (2002) J Am Chem Soc 124:14580–14585
8. Holmlin RE, Dandliker PJ, Barton JK (1997) Angew Chem Int Ed Engl 36:2714–2730
9. Williams TT, Odom DT, Barton JK (2000) J Am Chem Soc 122:9048–9049
10. Schuster GB (2000) Acc Chem Res 33:253–260
11. Giese B (2000) Acc Chem Res 33:631–636
12. Giese B, Spichty M (2000) Chem Phys Chem 1:195–198
13. Lewis FD, Wu T, Letsinger RL, Wasielewski MR (2001) Acc Chem Res 34:159–170
14. Grinstaff MW (1999) Angew Chem Int Ed 38:3629–3635
15. Saito I, Nakamura T, Nakatani K, Yoshioka Y, Yamaguchi K, Sugiyama H (1998) J Am Chem Soc 120:12686–12687
16. Sugiyama H, Saito I (1996) J Am Chem Soc 118:7063–7068
17. Voityuk AA, Jortner J, Bixon M, Rösch N (2000) Chem Phys Lett 324:430–434
18. Nakatani K, Dohno C, Saito I (1999) J Org Chem 64:6306–6311
19. Beratan DN, Priyadarshy S, Risser SM (1997) Chem Biol 4:38
20. Bixon M, Jortner J (1997) J Chem Phys 107:5154–5170
21. Jortner J, Bixon M, Langenbacher T, Michel-Beyerle ME (1998) Proc Natl Acad Sci USA 95:12759–12765
22. Bixon M, Giese B, Wessely S, Langenbacher T, Michel-Beyerle ME, Jortner J (1999) Proc Natl Acad Sci USA 96:11713–11716
23. Yoshioka Y, Kitagawa Y, Takano Y, Yamaguchi K, Nakamura T, Saito I (1999) J Am Chem Soc 121:8712–8719
24. Bixon M, Jortner J (2001) J Am Chem Soc 123:12556–12567
25. Luger K, Mäder AW, Richmond RK, Sargent DF, Richmond TJ (1997) Nature 389:251–260
26. Rajski SR, Barton JK (2001) Biochemistry 40:5556–5564
27. Núñez ME, Noyes KT, Barton JK (2002) Chem Biol 9:403–415
28. Newman M, Strzelecka T, Dorner LF, Schildkraut I, Aggarwal AK (1995) Science 269:656–663
29. Giese B, Spichty M (2000) Chem Phys Chem 1:195–198
30. Lewis FD, Kalgutkar RS, Wu Y, Liu X, Liu J, Hayes RT, Miller SE, Wasielewski MR (2000) J Am Chem Soc 122:12346–12351
31. Qin X-Z, Williams F (1987) J Am Chem Soc 109:595-597
32. Kim J-M, Bogdan MA, Mariano PS (1991) J Am Chem Soc 113:9251–9257
33. Bouchoux G, Alcaraz C, Dutuit O, Nguyen MT (1998) J Am Chem Soc 120:152–160
34. Zhong B, Silverman RB (1997) J Am Chem Soc 119:6690–6691
35. Lee J, U JS, Blackstock SC, Cha JK (1997) J Am Chem Soc 119:10241–10242
36. Shaffer CL, Morton MD, Hanzlik RP (2001) J Am Chem Soc 123:349–350
37. Musa OM, Horner JH, Shahin H, Newcomb M (1996) J Am Chem Soc 118:3862–3868
38. (a) Acedo M, Fàbrega C, Aviño A, Goodman M, Fagan P, Wemmer D, Eritja R (1994) Nucleic Acid Res 22:2982–2989
39. Adib A, Potier PF, Doronina S, Huc I, Behr J-P (1997) Tetrahedron Lett 38:2989–2992
40. Dohno C, Ogawa A, Nakatani K, Saito I (2003) J Am Chem Soc 125:10154–10155.

Top Curr Chem (2004) 236:187–204
DOI 10.1007/b94416

Excess Electron Transfer in Defined Donor-Nucleobase and Donor-DNA-Acceptor Systems

Christoph Behrens · Michaela K. Cichon · Friederike Grolle · Ulrich Hennecke · Thomas Carell

Fachbereich Chemie, Philipps-Universität Marburg, Hans-Meerwein-Strasse, 35032 Marburg, Germany
E-mail: carell@mailer.uni-marburg.de

Abstract The transfer of positive charge through DNA has been investigated in great detail over the last couple of years. In this area, major new mechanistic insights have been gained using defined acceptor modified DNA strands. Transport of exon electrons, in contrast, is much less well-explored. Our current mechanistic understanding is based on EPR spectroscopic studies of DNA material reduced using solvated electrons. Herein we report the development of defined donor-acceptor modified DNA double strands which allow the study of excess electron transfer with high precision. The model mimics the DNA repair process of DNA photolyases: they contain a reduced and deprotonated flavin as a light-triggered electron donor and a thymine dimer as the electron acceptor. The dimer performs a cycloreversion upon single electron reduction, which translates the electron capture event into a readily detectable strand break signal. Investigations with these model systems allowed us to clarify that electrons hop through DNA using pyrimidine bases as stepping stones.

1 Introduction

The transfer of a positive charge – a hole – through DNA is of fundamental importance for the process of DNA damage formation [1]. Single electron oxidation of DNA, from the reaction of a reactive oxygen species or a photochemically triggered electron acceptor with DNA for example, yields a guanine radical cation in close vicinity to the initial oxidation site. The radical cation always localises at guanines because guanine is the nucleobase with the lowest oxidation potential [2, 3]. γ-Radiolysis studies [4, 5] and investigations of defined electron acceptor DNA systems have revealed over the past decade that the formed guanine radical cation participates in long range charge delocalisation processes (Fig. 1) [6–9]. It has been found that an adjacent guanine in the DNA duplex is able to donate an electron to the initial guanine radical cation site. The sequential electron transfer between guanines allows rapid delocalisation of the positive charge in DNA. The charge virtually hops through DNA using guanines as stepping stones [10, 11]. Although this hopping process is now well accepted, particularly due to the fundamental contribution by J.K. Barton [6], B. Giese [7], F.D. Lewis [12], G. Schuster [13] and M.R. Wasielewski [14], the mechanistic details of the long range hole hopping process are still under intensive investigation and certain observations are still controversially debated [15–18].

All researchers in the field, however, agree that the guanine radical cations react with water or oxygen to give oxidative DNA lesions, which are in vivo

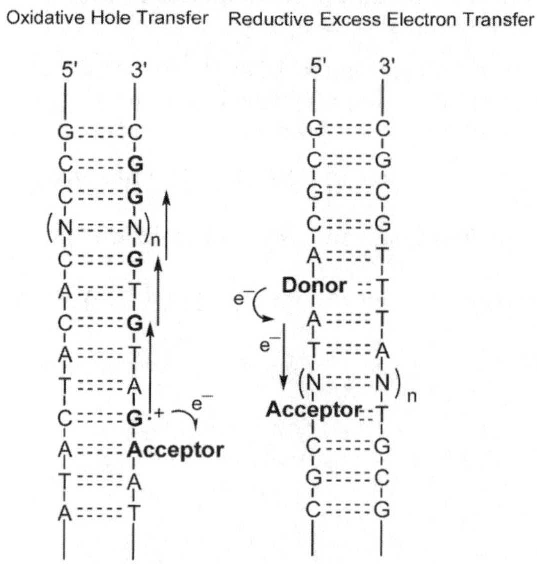

Fig. 1 Oxidative transfer of a radical cation through DNA and reductive transfer of an excess electron through DNA

mutagenic and responsible for cell death [19]. These oxidative lesions require efficient cellular DNA repair in order to avoid the harmful effects of oxidative DNA damage [20]. Although our knowledge about oxidative hole transfer and its biological consequences have developed over the last years to a point where we now have a rather sound understanding of the basic underlying processes, rather less is known about the transfer of excess electrons – negative charges – through DNA, as depicted in Fig. 1 [12, 21]. If an electron donor injects electrons into a duplex, are these negative charges also able to move, maybe by hopping, to an acceptor site in DNA [22]? Are there any natural processes known which utilize single electron reduction of nucleobases? These are the questions that we address in this review article by outlining recent results obtained with defined synthetic model systems. Answers to these questions are believed to be the basis for the development of DNA as a charge transport medium [23, 24] in nanoelectronic devices [25, 26].

2 Single Electron Reduction of Nucleobases in Biology and in Model Systems

2.1 Single Electron Transfer Based DNA Repair

Electron donation to nucleobases is a fundamental process exploited by nature to achieve the efficient repair of UV induced lesions in DNA [27, 28]. Nature developed to this end two enzymes, CPD photolyases and (6–4) photolyases, which both inject electrons into the UV damaged DNA bases [29, 30]. Both enzymes are, in many species, including plants, essential for the repair of the UV-light induced DNA lesions depicted in Scheme 1 [31].

UV-irradiation of cells induces the formation of two major photo products, which are mutagenic and responsible for UV induced cell death (Scheme 1). The first, cyclobutane pyrimidine dimers (T=T), are formed in a $[2\pi+2\pi]$ cycloaddition reaction in response to the absorption of a UV photon by a pyrimidine base in a dipyrimidine sequence like TpT. The second, (6–4)-photo products, are generated after a Paternó-Büchi reaction of two pyrimidines followed by ring opening of the oxetane intermediate, which is unstable above −80°C (Scheme 1). CPD photolyases recognize and exclusively repair cyclobutane pyrimidine dimer lesions in DNA. (6–4)-photolyases repair (6–4)-lesions. Both repair processes are light triggered as depicted in Scheme 2 [30]. So far, only the mechanism of the CPD repair has been deciphered in detail. The sequence and cofactor homology between CPD and (6–4)-photolyases, however, suggest a very similar mechanism.

The light energy is initially absorbed by an antenna pigment in both photolyases. Either a 8-hydroxy-5-deazaflavin (8-HDF) or a methenyltetrahydrofolate (MTHF) functions as the absorbing pigment, depending on the type of photolyase. The energy is then transferred to a reduced flavin coenzyme ($FADH^-$), which donates an electron to the CPD lesion. Upon single electron

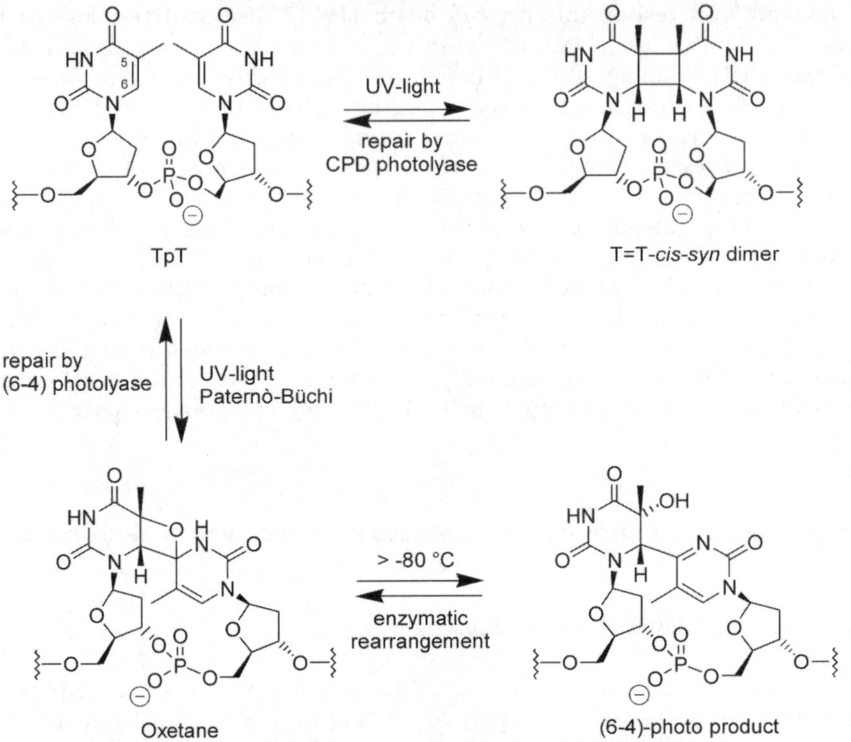

Scheme 1 UV-light induced formation of the two major photo lesions in DNA. T=T: cyclobutane pyrimidine dimer. (6–4)-photo product: (6–4)-lesion, formed after ring opening of an oxetane intermediate, which is the product of a Paternó-Büchi reaction

reduction, the dimer spontaneously cleaves into the two monomers (repair). The closely related (6–4)-photolyase seems to repair (6–4)-lesions in a very similar process. This enzyme is thought to re-isomerise the (6–4)-lesion back into the oxetane [32], which is again then split by single electron reduction from the reduced FAD-coenzyme. The mechanism of the rearrangement reaction to the oxetane is currently under intensive investigation.

2.2 Single Electron Reduction of Nucleobases and UV-Lesions

Although the reduction potentials of DNA bases and UV induced DNA lesions inside a DNA double strand or inside the active site of a DNA photolyase, together with the reduction potential of the photoexcited $FADH^{-*}$ in the photolyases, are not known, currently available redox potentials indicate that the single electron reduction of a nucleobase or a UV induced dimer lesion by a reduced and deprotonated flavin coenzyme is a weakly exothermic process. The reduced and deprotonated $FADH^{-*}$ in its photoexcited state is

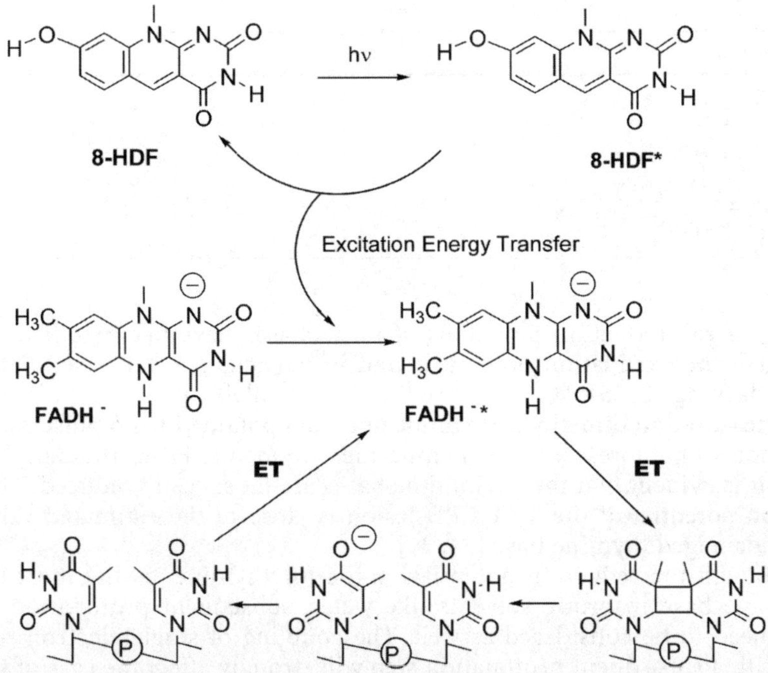

Scheme 2 Mechanism of repair of cyclobutane pyrimidine dimers (CPD) by a CPD photolyase. 8-HDF: 8-hydroxy-5-deazaflavin, ET: electron transfer. $FADH^{-}$: reduced and deprotonated flavin-coenzyme

believed to have a reduction potential between $E_{red*}= -2.6$ V and -2.8 V [33, 34]. The reduced $FADH^{-*}$ in its excited state is therefore an extremely strong electron donor, able to reduce all nucleobases and also CPD lesions. The currently available reduction potentials are listed in Table 1.

Table 1 Reduction potentials of some nucleobases

Base	Reduction potentials in V		
	E(Red)[a]	E(Red)[b]	E(Red)[c]
dG	−2.76		
dA	−2.45		
dC	−2.23	−1.09	−2.16 (DMC)
dT	−2.14	−1.10	−2.14 (DMT)
U	−2.04	−1.10	−2.11 (DMU)
T=T			−2.20 (DMTD)

DMC=Dimethylcytosine, DMT=Dimethylthymine, DMU=Dimethyluridine, DMTD=Dimethylthymine-Dimer

[a] Polarographic potentials in DMF versus NHE [2]

[b] Data from pulse radiolysis experiments in water at pH=8.5 against NHE [35]

[c] Data from fluorescence quenching experiments in acetonitrile against SCE [34, 36]

Table 2 pK_a-values of pyrimidine nucleobases [37]

Equilibrium	pK_a
$TH^{\cdot} \rightleftharpoons T^{\cdot\ominus} + H^{\oplus}$	6.9
$TH^{\oplus} \rightleftharpoons T + H^{\oplus}$	−5
$AH \rightleftharpoons A^{\ominus} + H^{\oplus}$	>14
$CH^{\cdot} \rightleftharpoons C^{\cdot\ominus} + H^{\oplus}$	13
$CH^{\oplus} \rightleftharpoons C + H^{\oplus}$	4.3
$GH \rightleftharpoons G^{\ominus} + H^{\oplus}$	9.5

In general, reduction potentials of nucleobases have been studied much less than their oxidation potentials, and in particular water-based data are rather lacking [2, 35]. We therefore listed the available polarographic potentials measured in dimethylformamide and data obtained from pulse radiolysis studies or fluorescence quenching measurements. From the data in Table 1, it is evident that the pyrimidine bases are most easily reduced. The reduction potential of the T=T CPD lesion is close to the estimated value of the undamaged thymine base [34, 36].

Although the reduction potentials argue for thymine, as the most easily reducable base in protic solvents like water, subsequent protonation reactions need to be considered as well. The coupling of single electron reduction with a subsequent protonation step will strongly affect the ease of single electron reduction. Table 2 contains the pK_a-values of some nucleobases in their reduced and neutral states [37]. It is clear that the thymine radical anion, due to its rather neutral pK_a-value of about 7 is unlikely to become protonated either by water or by the adenine counter base in the DNA strand.

The high pK_a-value of the cytosine radical anion, however, will enforce rapid protonation in water and maybe also by the guanine counterbase in a DNA double strand, as depicted in Scheme 3 [38]. In fact, it is currently be-

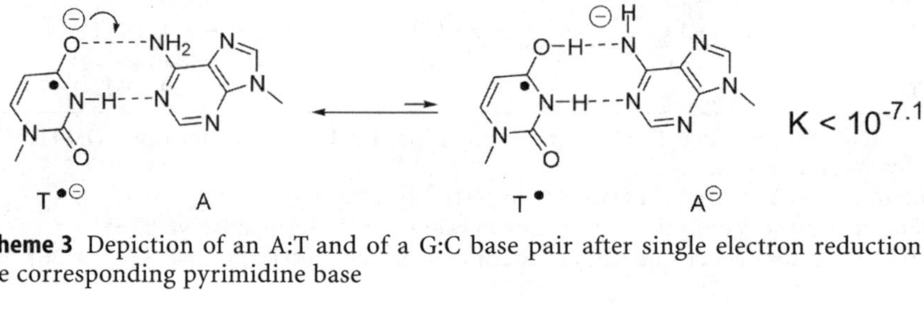

Scheme 3 Depiction of an A:T and of a G:C base pair after single electron reduction of the corresponding pyrimidine base

lieved that the protonation of the cytosine radical anion shifts the reduction potential to a more positive value, so that reduction of thymine and cytosine in DNA may be almost equally efficient. In fact, at temperatures below 77 K, cytosine seems to be the trap for excess electrons in DNA. These protonation reactions are reversible, which may allow a thermally activated proton-gated electron transfer at higher temperature [39]. Irreversible protonation at C-6 and C-5 of the thymine and cytosine radical anions may also occur, which would interfere with the charge hopping process [39].

Base stacking will further modulate the reduction potentials. Initial calculations show that pyrimidines flanked by other pyrimidines, like in 3′-TTT-5′ or 3′-TCT-5′ sequences, are most easily reduced. Incorporation of a purine base close to the pyrimidine, however, seems to make the pyrimidine reduction more difficult [40].

2.3
Single Electron Transfer in Model Compounds

In order to investigate the single electron donation process from a reduced flavin to a pyrimidine dimer or oxetane lesion, the photolyase model compounds **1–4** depicted in Scheme 4 were prepared [41, 42]. The first model compounds **1** and **2** contain a cyclobutane uracil (**1**) or thymine (**2**) dimer covalently connected to a flavin, which is the active electron donating subunit in photolyases. These model systems were dissolved in various solvents

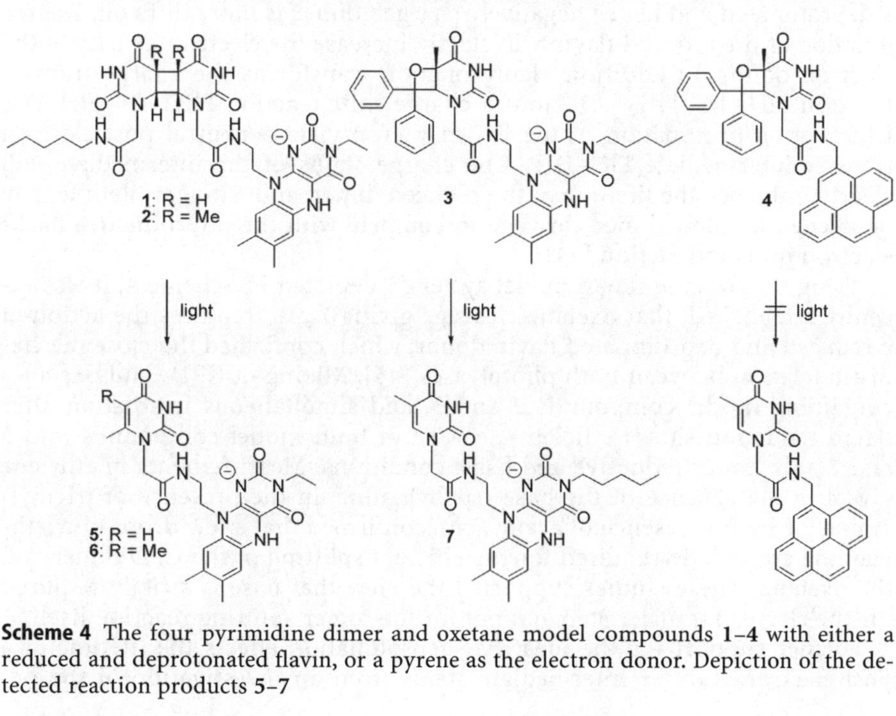

Scheme 4 The four pyrimidine dimer and oxetane model compounds **1–4** with either a reduced and deprotonated flavin, or a pyrene as the electron donor. Depiction of the detected reaction products **5–7**

such as water, ethanol, dioxane or acetonitrile and placed into sealed cuvettes. Reduction of the flavin under strictly anaerobic conditions allowed us to prepare stable solutions of the model compounds **1** and **2** containing the flavin in the reduced, and hence active, electron donating state. Irradiation of these solutions with daylight or monochromatic light of λ=360 nm, and analysis of the assay solution by reversed phase HPLC showed rapid light dependent cleavage of the cyclobutane pyrimidine dimer units **1** and **2** to give the products **5** and **6** as shown in Scheme 4 [42].

These experiments proved that a light-excited, reduced flavin is indeed able to photoreduce cyclobutane pyrimidine dimers and that these dimers undergo a spontaneous cycloreversion. The quantum yield of about Φ=5% clarified that the overall dimer splitting process is highly efficient, even in these simple model systems (ϕ photolyase $\approx$70%).

Investigation of the pH-dependency of the splitting reaction was required in order to investigate how the protonation state of the reduced flavin affects the electron transfer process and hence the cleavage reaction [42, 43]. Measurements of the model compounds in water buffered at various pH-values showed maximal cleavage efficiency between pH=7 and 9. No dimer cleavage was observed above pH=10 and at or below pH=5. These values match with the pK_a-values of the reduced flavin (p$K_a \approx$6.5) and of the cyclobutane pyrimidine dimer (p$K_a \approx$11). Therefore the data provided clear proof that deprotonation of the reduced flavin is strictly required for the electron transfer driven cleavage process. In contrast, deprotonation of the dimer unit at high pH values stops the repair reaction, possibly because the electron transfer to a deprotonated and hence negatively charged dimer is more difficult. Deprotonation of the reduced flavin will clearly increase the electron density of the electron donor. In addition, deprotonation transforms the charge transfer reaction $FlH_2\text{-}D \rightarrow FlH_2^{\cdot+}\text{-}D^{\cdot-}$ into a charge shift reaction $FlH^-\text{-}D \rightarrow FlH^{\cdot}\text{-}D^{\cdot-}$, which upon protonation of the $D^{\cdot-}$ may even yield a neutral post-electron transfer intermediate $FlH^{\cdot}\text{-}DH^{\cdot}$. The charge status of the intermediate will clearly influence the lifetime of the reduced dimer, and a longer lifetime may be needed to allow dimer cleavage to compete with the unproductive back-electron transfer reaction [44].

Using the oxetane-flavin model system **3** depicted in Scheme 4, it was recently established, that oxetane cleavage giving **7** also requires the action of a reduced and deprotonated flavin donor, which confirmed the close mechanistic relation between both photolyases [45]. Mixing of CPD- and oxetane-containing model compounds 2 and 3, and simultaneous irradiation after flavin reduction shows efficient cleavage of both model compounds into **6** and **7** only under reductive and basic conditions. Measurements in ethylene glycol in the absence of the base triethylamine, in the presence of triethylamine, or in the presence of acetic acid, confirmed that a few drops of triethylamine are strictly required for the efficient splitting of the CPD dimer and the oxetane. These studies supported the view that base is strictly required for the electron transfer step and not for the dimer splitting reaction itself.

Further support for the idea that deprotonation effects the lifetime of a post-electron transfer intermediate stems from an investigation of the pyrene-oxetane model compound **4** (Scheme 4). The pyrene is clearly able to function, after photoexcitation, as a strong single electron donor in charge transfer reactions. The pyrene is potent enough to reduce pyrimidines, cyclobutane pyrimidine dimers and also oxetanes to give a $Py^{\cdot +}$-$D^{\cdot -}$ intermediate [46]. Irradiation of the pyrene model compound **4**, however, gave under no circumstances dimer cleavage.

Whatever the reason may be behind the strict necessity to deprotonate the flavin donor, the reduced and deprotonated flavin was established in these model studies to be an efficient electron donor, able to reduce nucleobases and oxetanes. In the model compounds **1** and **2** the pyrimidine dimer translates the electron transfer step into a rapidly detectable chemical cycloreversion reaction [47, 48]. Incorporation of a flavin and of a cyclobutane pyrimidine dimer into DNA double strands was consequently performed in order to analyse the reductive electron transfer properties of DNA.

3
Excess Electron Transfer in DNA:DNA Double Strands

Incorporation of a flavin electron donor and a thymine dimer acceptor into DNA double strands was achieved as depicted in Scheme 5 using a complex phosphoramidite/*H*-phosphonate/phosphoramidite DNA synthesis protocol. For the preparation of a flavin-base, which fits well into a DNA double strand structure, riboflavin was reacted with benzaldehyde-dimethylacetale to rigidify the ribityl-chain as a part of a 1,3-dioxane substructure [49]. The benzacetal-protected flavin was finally converted into the 5′-dimethoxytrityl-protected-3′-*H*-phosphonate ready for the incorporation into DNA using machine assisted DNA synthesis (Scheme 5a). For the cyclobutane pyrimidine dimer acceptor, a formacetal-linked thymine dimer phosphoramidite was prepared, which was found to be accessible in large quantities [50]. Both the flavin base and the formacetal-linked thymidine dimer, were finally incorporated into DNA strands like **7–12** (Scheme 5c). As depicted in

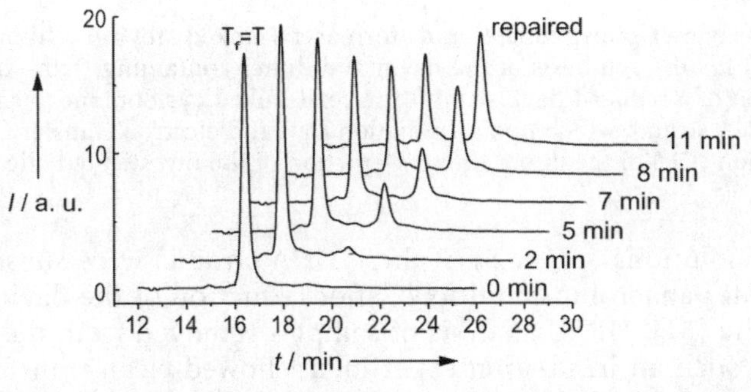

Fig. 2 Reversed phase HPLC analysis of a solution containing the oligonucleotide 7 during irradiation (0–11 min). The dimer containing DNA strand elutes at a retention time of about 16.5 min. The DNA strand after cycloreversion of the dimer elutes at about 19 min. Clearly evident is the clean formation of the product DNA strand upon irradiation

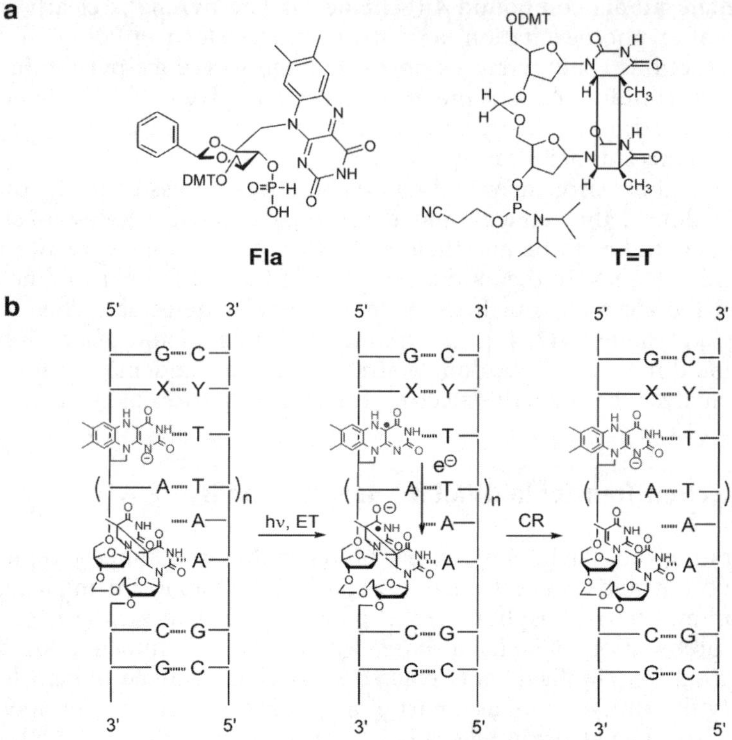

c

7	5'-d(GCGCA-**Fla-T=T**AACGC)-3' 3'-d(CGCGT- T -A ATTGCG)-5'	**8**	5'-d(GCGCA-**Fla-T=T**AACGC)-3' 3'-d(CGCGT- C -A ATTGCG)-5'
9	5'-d(GCGCA-**Fla-T=T**AACGC)-3' 3'-d(CGCGT- G -A ATTGCG)-5'	**10**	5'-d(GCGCA-**Fla-T=T**AACGC)-3' 3'-d(CGCGT- A -A ATTGCG)-5'
11	5'-d(GCGCA-**Fla-AT=T**ACGC)-3' 3'-d(CGCGT- T -TA ATGCG)-5'	**12**	5'-d(GCGCA-**Fla-AT=T**ACGC)-3' 3'-d(CGCGT- C -TA ATGCG)-5'

Scheme 5 **a** Flavin-*H*-phosphonate and formacetal-linked thymine dimer phosphoramidite used for the synthesis of the flavin and dimer containing DNA-strands **7–12**. **b** Representation of a reduced flavin- and formacetal-linked cyclobutane pyrimidine dimer containing DNA strand, which upon irradiation (hν) and electron transfer (ET) performs a cycloreversion (CR) of the dimer unit. **c** Depiction of the investigated oligonucleotides

Scheme 5b, solutions (pH=7.5) of these DNA strands were subsequently irradiated under anaerobic conditions, after reduction of the flavin with sodium dithionite [51]. HPLC analysis of samples removed from the assay solution during such an irradiation experiment showed clean transformation of the T_f=T-containing DNA strands into the repaired T_fT-containing DNA strands (Fig. 2). These experiments showed for the first time that a reduced flavin, if incorporated into DNA, can inject an extra electron into a DNA duplex. This electron is successfully captured by the cyclobutane pyrimidine dimer. The dimer performs in response to single electron reduction a cycloreversion within the oligonucleotide environment.

The cycloreversion experiments showed a clean T_f=T-DNA to T_fT-DNA transformation. No by-products were detected, which supports the idea that DNA may be more stable towards reduction compared to oxidation. Even heating the irradiated DNA with piperidine furnished no other DNA strands other then the repaired strands, showing that base labile sites – indicative for DNA damage – are not formed in the reductive regime. The quantum yield of the intra-DNA repair reaction was therefore calculated based on the assumption that the irradiation of the flavin-T_f=T-DNA strands induces a clean intramolecular excess electron transfer driven cycloreversion. The quantum yield was found to be around Φ=0.005, which is high for a photoreaction in DNA. A first insight into how DNA is able to mediate the excess electron transfer was gained with the double strands **11** and **12** in which an additional A:T base pair compared to 7 and 8 separates the dimer and the flavin unit.

Investigation of these double strands **11** and **12** gave the surprising result that repair in these strands was only by a factor of two less efficient compared to **7** and **8**, although the distance was increased from 3.4 Å to 6.8 Å.

For a more detailed investigation of excess electron transfer over larger distances it became apparent that the analytical detection of the ring-opening reaction of the thymine dimer would be a fundamental problem. Particularly in long oligonucleotides the available analytical techniques like HPLC or capillary electrophoresis failed to detect dimer reversal. This problem was solved with the development of the silyl-spaced thymine dimer depicted in Scheme 6a [52]. Incorporation of this dimer building block into flavin-containing DNA, and careful treatment of the DNA with fluoride, allowed us to cleave the silyl-bridge in the dimer, while leaving the sensitive flavin electron donor intact. Using the silyl-dimer and the fluoride treatment, DNA strands were prepared which contained beside the flavin, a cyclobutane dimer with an opened backbone, as depicted in Scheme 6b. Irradiation of solutions containing these DNA strands under the usual anaerobic conditions, after reduction of the flavin with dithionite, initiated again the electron transfer (ET) reaction, which now prompted a clean strand break. This allowed us to translate the excess electron transfer event in DNA into a DNA cleavage process, readily detectable by gel electrophoresis or ion exchange HPLC (Scheme 6b). With this technology in hand, analysis of the long range excess electron transfer properties of DNA was possible. The investigation was performed with the five DNA-strands **13–17** depicted in Scheme 6c [53].

Figure 3 shows an ion exchange HPLC analysis of the irradiation experiment performed with the duplex **13**. The starting duplex **13** elutes at a retention time of 13 min. The cleaved DNA strand has a retention time of 11 min. Clearly evident again is the clean conversion of the unrepaired duplex **13** into the cleaved, "repaired", DNA fragment.

In strand **13**, the flavin donor and the dimer acceptor are adjacent to each other, located at a distance of 3.4 Å. In the other four strands **14–17**, the dis-

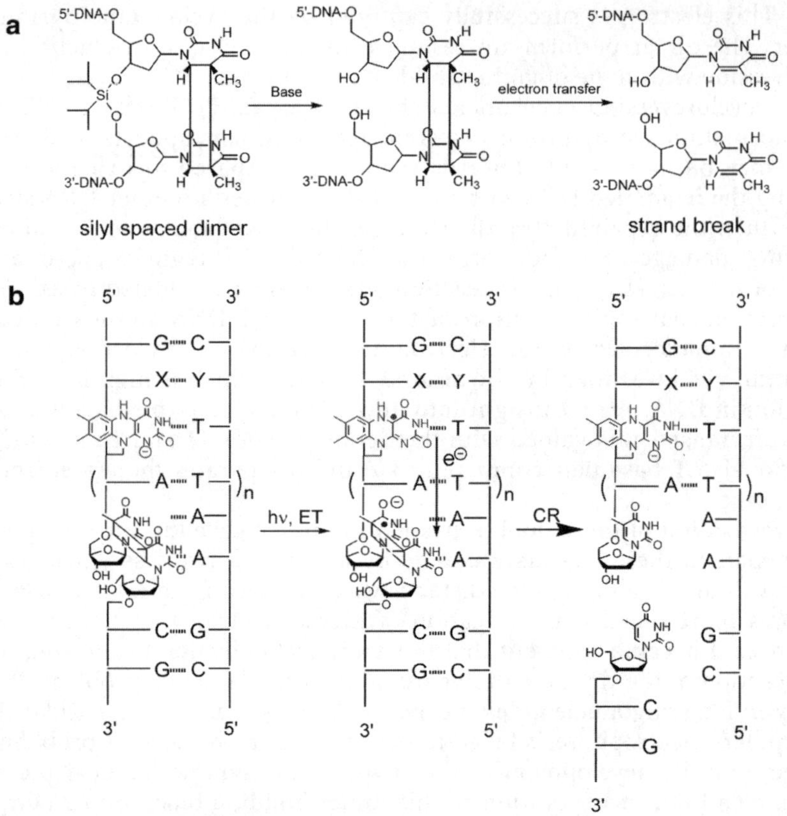

c		
13	5'-d(GCGCA-**Fla-T=T**AACGCG)-3' 3'-d(CGCGT T A ATTGCGC)-5'	Distance = 3.4 Å 6% Repair per min.
14	5'-d(GCGC-**Fla**-AAA**T=T**CGCG)-3' 3'-d(CGCG T TTTA AGCGC)-5'	Distance = 13.6 Å 2.2% Repair per min.
15	5'-d(GCGC-**Fla**-AAAA**T=T**CGCG)-3' 3'-d(CGCG T TTTTA AGCGC)-5'	Distance = 17 Å 1.8% Repair per min.
16	5'-d(GCGC-**Fla**-AAAAA**T=T**CGCG)-3' 3'-d(CGCG T TTTTTA AGCGC)-5'	Distance = 20.4 Å 1.0% Repair per min.
17	5'-d(GCGC-**Fla**-AAAAAA**T=T**CGCG)-3' 3'-d(CGCG A TTTTTTA AGCGC)-5'	Distance = 23.8 Å 0.8% Repair per min.

Scheme 6 a Incorporation of a silyl-spaced dimer into DNA, and treatment of the DNA strand with fluoride furnished DNA strand containing a dimer-unit with an open backbone. **b** Schematic representation of the light (ho) induced excess electron transfer (ET) from the reduced flavin to the dimer, followed by cycloreversion (CR). **c** The five DNA duplexes **13–17** containing the flavin donor and the dimer acceptor at increasing distances, together with the measured repair yields after irradiation for 1 min

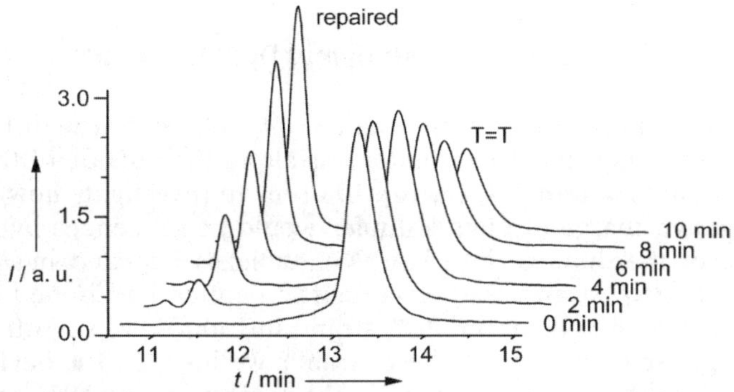

Fig. 3 Ion exchange HPLC analysis of an irradiation experiment performed with the DNA duplex **13** over a total irradiation time of 10 min. Clearly evident is the clean conversion of the duplex **13** into the cleaved duplex

tance is increased to 13 Å (**14**), 17 Å (**15**), 20.4 Å (**16**) and 23.8 Å (**17**). The distance increase was achieved by incorporation of an increasing number of A:T base pairs between the flavin and the dimer. Analysis of the cleavage efficiency under identical illumination conditions (360±5 nm) revealed 2.2% repair per minute in the double strand **14**. This value decreased to 1.8% repair per minute in **15**, to about 1.0% repair per minute in **16**, and to a final value of 0.8% per minute in **17**. These data show that an excess electron can indeed efficiently travel over a distance of at least 24 Å to repair a thymine dimer reductively at a remote site [53]. Up until this result, remote thymine dimer repair was only observed in the oxidative "mode" involving long range hopping of a hole to a thymine dimer by the Barton group [18, 54, 55]. It is still under debate to whether the oxidative remote thymine dimer repair is more or less efficient than the competing guanine radical cation formation, followed by the reaction of this nucleobase radical cation with water or oxygen to give oxidative DNA lesions [17].

The rather small decrease of the repair efficiency with increasing distance indicates that the excess electron might not travel by a Marcus mechanism. The Marcus model predicts a decrease of the cleavage efficiency by a factor between about 7 and 10 with every additional A:T base pair introduced between the dimer and the flavin [56, 57]. The observed decrease by just a factor of less than two in our experiment points to an excess electron hopping mechanism, in which the electron hops via intermediate thymines, which in such a scenario function as stepping stones along the electron transfer pathway. Our result therefore supports radiolysis and EPR data, which suggest that an excess electron can hop through DNA above a temperature of −80 °C [5, 58–60].

4
Excess Electron Transfer in DNA:PNA Hybrid Double Strands

Recently it was speculated that the transfer of a charge through DNA is accompanied by a movement of counterions along the outside of the polyanionic DNA double strand [13, 15, 16]. In order to investigate how the transfer of a negative charge through a duplex would be influenced by charges at the outside of the duplex, the DNA:PNA duplexes depicted in Fig. 4 were prepared. The PNA strand features a neutral peptide backbone [61]. Lysin-residues (L) at both ends of the PNA strand introduce even positive charges at the N- and the C-termini. In these strands we inserted the flavin into the PNA part, while the dimer remained in the corresponding DNA strand. The excess electron has to hop consequently in these constructs in an intermolecular fashion from the PNA strand to the opposite DNA strand in order to reach the dimer. The DNA:PNA strands (Fig. 4) were irradiated, again under anaerobic conditions after reduction of the flavin with dithionite.

Analysis of the assay solution by ion exchange HPLC showed efficient dimer repair, with quantum yields (Φ=0.002) very similar to those obtained for the DNA:DNA situation. These studies confirmed that an excess electron transfer proceeds even between strands and that changing one polyanionic strand into a neutral strand does not change the electron transfer efficiency dramatically. Using the DNA:PNA hybrid strands with the rather large distance between the flavin and the dimer, it was also shown that the excess electron can move even through these hybrid duplexes over a relatively large distance [62]. In duplex **P2**, the distance between the **F1** and the **T=T** is increased from 3.4 Å (**P1**) to about 17 Å (**P2**). In **P2**, the **T=T** cleavage is by a factor of about 12 less efficient, showing that every additional base pair between the two redox partners on averages reduces the cleavage efficiency again only by a factor of about 2. We next explored the distance dependence of the electron transfer process with the two double strands **P4** and **P6**. Here again the distance is increased from 3.4 Å (**P4**) to 17 Å (**P6**). In **P4** and **P6**, the distance effect is, however, much more pronounced. The splitting efficiency is reduced in **P6** by a factor of about 100 and is in consequence only slightly above background. The much stronger distance dependence was confirmed with the double strand **P5**, in which the cleavage efficiency is reduced already by a factor of 30, although the distance is only increased to about 10 Å realtive to **P4**.

The fact that the stacking situation also influences the excess electron transfer was finally established with the double strands **P3** and **P7**. **P3** exhibits increased fraying of the **T=T** containing double strand ends due to a lack of proper base pairing. Strand **P7** contains, instead of **F1**, the flavin donor **F2**, which is located more deeply inside the duplex. The transfer efficiency responds to both changes. The excess electron transfer is decreased in **P3** compared to **P1** and increased in **P7** relative to **P5**.

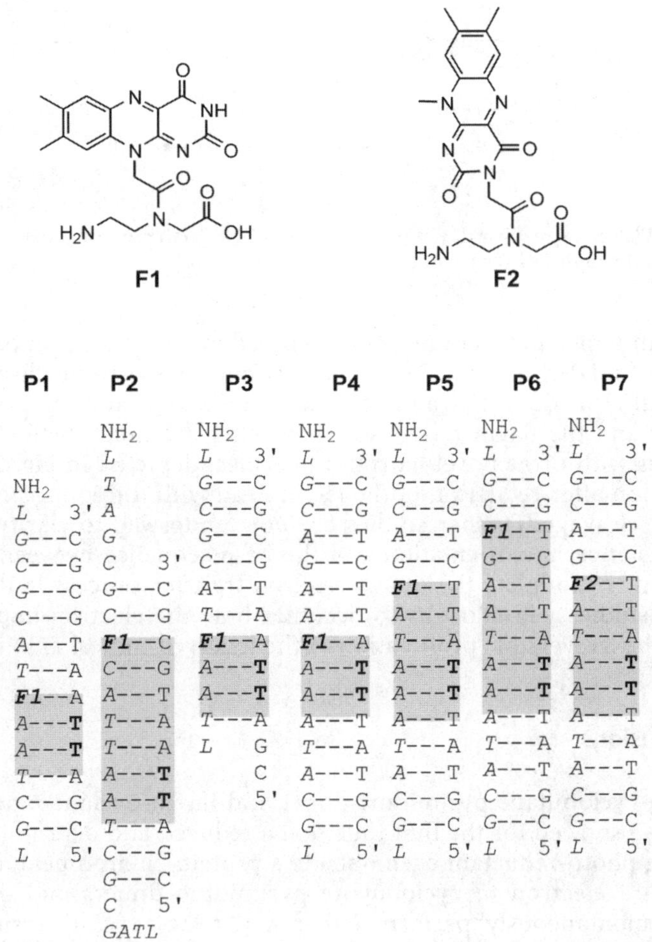

	P1	P2	P3	P4	P5	P6	P7
M.p. (+3°):	52	55	(30)[a]	45	42	47	40
Φ_{rel} (+ 20%):	1	0.08	0.6				
Φ_{rel} (+ 20%):				1	0.03	0.01	0.05

Fig. 4 Flavin amino acids **F1** and **F2** used for the preparation of DNA:PNA hybrid duplexes. These duplexes possess the flavin in the PNA strand and the "backbone" opened dimer in the DNA strand. [a] Indicates unstructured melting. M.p.= melting points of the constructs

5
Excess Electron Transfer in Flavin-Capped DNA Hairpins

In order to gain a more detailed insight into how the flavin and thymine dimer orientation may influence the excess electron transfer process, rigid

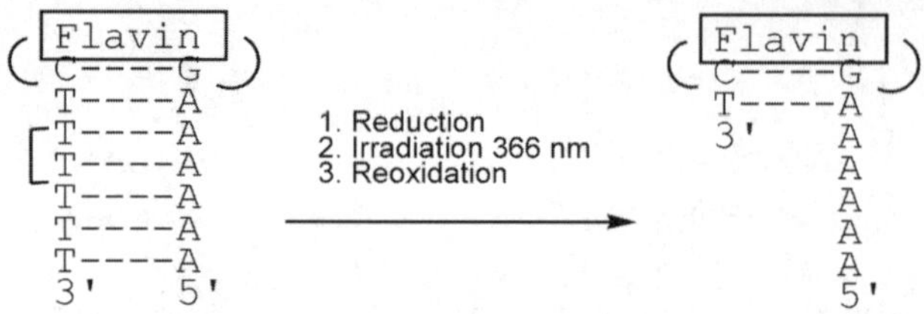

Fig. 5 Flavin-capped and dimer containing DNA hairpin for excess electron transfer studies through DNA

hairpin molecules were recently prepared in which a flavin bridge forms the head of a DNA hairpin. These hairpin molecules are much more rigid than normal DNA double strands, so that the distance and the orientation of the dimer and the flavin relative to each other are much better defined. Initial studies with these novel hairpin molecules, depicted in Fig. 5, show that irradiation after reduction of the flavin head with dithionite provides efficient dimer cleavage. Further studies are now underway to clarify precisely how the distance, the orientation and the base sequence between the flavin and the dimer modulate the excess electron transfer process [63]. We hope that these studies will allow us to decipher how the electron hopping competes with the irreversible protonation of the involved radical anions.

6 Conclusions

Flavin-cyclobutane pyrimidine dimer and flavin-oxetane model compounds like **1**–**3** showed for the first time that a reduced and deprotonated flavin is a strong photo-reductant even outside a protein environment, able to transfer an extra electron to cyclobutane pyrimidine dimers and oxetanes. There then spontaneously perform either a $[2\pi+2\pi]$ cycloreversion or a retro-Paternó-Büchi reaction. In this sense, the model compounds mimic the electron transfer driven DNA repair process of CPD- and (6–4)-photolyases.

Incorporation of an artificial flavin nucleobase and of a cyclobutane pyrimidine dimer building block into DNA:DNA double strands, DNA:PNA hybrid duplexes, and DNA-hairpins, provided compelling evidence that an excess electron can hop through DNA to initiate dimer repair even at a remote site. The maximum excess electron transfer distance realised so far in these defined Donor-DNA-Acceptor systems is 24 Å. New experiments are now in progress to clarify whether even larger transfer distances can be achieved.

Because DNA is much more stable under reductive conditions, excess electron transfer may in the future, open the door to side reaction free chemistry at a distance, and may pave the way for DNA-based nanoelectronics devices.

References

1. Pouget JP, Douki T, Richard MJ, Cadet J (2000) Chem Res Toxicol 13:541
2. Seidel CAM, Schulz A, Sauer MHM (1996) J Phys Chem 100:5541
3. Steenken S, Jovanovic SV (1997) J Am Chem Soc 119:617
4. Melvin T, Botchway S, Parker AW, O'Neill P (1995) Chem Comm 653
5. Debije MG, Milano MT, Bernhard WA (1999) Angew Chem Int Ed 38:2752
6. Holmlin RE, Dandlicker PJ, Barton JK (1997) Angew Chem Int Ed 36:2715
7. Giese B (2000) Acc Chem Res 33:631
8. Lewis FD, Letsinger RL, Wasielewski MR (2001) Acc Chem Res 34:159
9. Schuster GB (2000) Acc Chem Res 33:253
10. Giese B (2000) Chem Phys Chem 1:195
11. Giese B, Amaudrut J, Köhler A-K, Spormann M, Wessely S (2001) Nature 4112:318
12. Lewis FD, Miller SE, Hayes RT, Wasielewski MR (2002) J Am Chem Soc 124:11280
13. Barnett RN, Cleveland CL, Joy A, Landman U, Schuster GB (2001) Science 294:567
14. Lewis FD, Wu T, Zhang Y, Letsinger RL, Greenfield SR, Wasielewski MR (1997) Science 277:673
15. Santosh U, Schuster GB (2002) J Am Chem Soc 124:10986
16. Williams TT, Barton JK (2002) J Am Chem Soc 124:1840
17. Dotse AK, Boone EK, Schuster GB (2000) J Am Chem Soc 122:6825
18. Vivic DA, Odom DT, Núñez ME, Gianolio DA, McLaughlin LW, Barton JK (2000) J Am Chem Soc 122:8603
19. Burrows CJ, Muller JG (1998) Chem Rev 98:1109
20. Friedberg EC, Walker GC, Siede W (1995) DNA repair and mutagenesis. ASM Press, Washington
21. Lewis FD, Liu X, Wu Y, Miller SE, Wasielewski MR, Letsinger RL, Sanishvili R, Joachimiak A, Tereshko V, Egli M (1999) J Am Chem Soc 121:9905
22. Giese B, Wessely S, Sporman M, Lindemann U, Meggers E, Michel-Beyerle ME (1999) Angew Chem Int Ed 38:996
23. Cai L, Tabata H, Kawai T (2000) Appl Phys Lett 77:3105
24. Yoo K-H-, Ha DH, Lee J-O, Park JW, Kim J, Kim JJ, Lee H-Y, Kawai T, Choi HY (2001) Phys Rev Lett 87:1981021
25. Fink H-W, Schönenberger C (1999) Nature 398:407
26. Porath D, Bezryadin A, De Vries S, Dekker C (2000) Nature 403:635
27. Heelis PF, Hartman RF, Rose SD (1995) Chem Soc Rev 24:289
28. Carell T (1995) Angew Chem Int Ed 34:2491
29. Sancar A (1994) Biochemistry 33:2
30. Carell T, Burgdorf LT, Kundu LM, Cichon MK (2001) Curr Op Chem Biol S 491
31. Kanai S, Kikuna R, Toh H, Ryo H, Todo T (1997) J Mol Evol 45:535
32. Hitomi K, Nakamura H, Kim S-T, Mizikoshi T, Ishikawa T, Iwai S, Todo T (2001) J Biol Chem 276:10103
33. Yeh S-R, Falvey DE (1992) J Am Chem Soc 114:7313
34. Scannel MP, Fenick DJ, Yeh S-R, Falvey DE (1997) J Am Chem Soc 119:1971
35. Steenken S, Telo JP, Novais HM, Candeias LP (1992) J Am Chem Soc 114:4701
36. Scannel MP, Prakash G, Falvey DE (1997) J Phys Chem A 101:4332
37. Steenken S (1997) Biol Chem 378:1293
38. Zhongli XL, Sevilla MD (2001) J Phys Chem B 105:10115
39. Debije MG, Bernhard WA (2002) J Phys Chem A 106:4608
40. Voityuk AA, Michel-Beyerle ME, Rösch N (2001) Chem Phys Lett 342:231
41. Carell T, Epple R, Gramlich V (1996) Angew Chem Int Ed 35:620
42. Epple R, Wallenborn E-U, Carell T (1997) J Am Chem Soc 119:7440
43. Hartman RF, Rose SD (1992) J Am Chem Soc 114:3559
44. Carell T, Epple R (1998) Eur J Org Chem 7:1245
45. Cichon MK, Arnold S, Carell T (2002) Angew Chem Int Ed 51:767

46. Wagenknecht H-A, Stemp EDA, Barton JK (2000) Biochemistry 39:5483
47. Epple R, Carell T (1998) Angew Chem Int Ed 37:938
48. Epple R, Carell T (1999) J Am Chem Soc 121:7318
49. Schwögler A, Carell T (2000) Org Lett 2:1415
50. Butenandt J, Eker APM, Carell T (1998) Chem Eur J 4:642
51. Schwögler A, Burgdorf LT, Carell T (2000) Angew Chem Int Ed 39:3918
52. Kundu LM, Burgdorf LT, Kleiner O, Batschauer A, Carell T (2002) Chem Bio Chem 3:1053
53. Behrens C, Burgdorf LT, Schwögler A, Carell T (2002) Angew Chem Int Ed 114:1841
54. Dandlicker PJ, Holmlin RE, Barton JK (1997) Science 275:1465
55. Dandlicker PJ, Núñez ME, Barton JK (1998) Biochemistry 37:6491
56. Bixon M, Jortner J (2001) J Am Chem Soc 123:12556
57. Jortner J, Bixon M, Langenbacher T, Michel-Beyerle ME (1998) Proc Natl Acad Sci USA 95:12759
58. Pezeshk A, Symons MCR, McClymont JD (1996) J Phys Chem 100:18562
59. Messer A, Carpenter K, Forzley K, Buchanan J, Yang S, Razskazovskii Y, Cai Z, Sevilla MD (2000) J Phys Chem B 104:1128
60. Cai Z, Gu Z, Sevilla MD (2000) J Phys Chem B 104:10406
61. Nielsen PE, Egholm M (1999) Horizon Scientific Press, Norfolk
62. Cichon MK, Haas CH, Grolle F, Mees A, Carell T (2002) J Am Chem Soc 124:13984
63. Behrens C, Ober M, Carell T (2002) Eur J Org Chem 3281

Author Index Volumes 201–236

Author Index Vols. 26–50 see Vol. 50
Author Index Vols. 51–100 see Vol. 100
Author Index Vols. 101–150 see Vol. 150
Author Index Vols. 151–200 see Vol. 200

The volume numbers are printed in italics

Subject Index

Printing: Saladruck, Berlin
Binding: Stein+Lehmann, Berlin